中国建筑工业出版社

交织 INTERWEAVE
上海南外滩地块城市设计　Urban Interworkings at Shanghai's South Bund

TONGJI UNIVERSITY 同济大学
Shanghai, China　　中国上海

RENSSELAER POLYTECHNIC INSTITUTE 伦斯勒理工学院
Troy, New York, United States 美国纽约

2

Acknowledgments
特别鸣谢

RPI LEADERSHIP 伦斯勒领导
PRESIDENT 校长 » Shirley Ann Jackson, Ph.D
DEAN 院长 » Evan Douglis

TONGJI LEADERSHIP 同济领导
PRESIDENT 校长 » Gang Pei, Ph.D 裴钢
DEAN 院长 » Changfu Wu 吴长福
DEPUTY DIRECTOR 副系主任 » Jianlong Zhang 张建龙

BOOK DESIGN 装帧设计
SENIOR EDITOR 高级编辑 » Gustavo Crembil
ASSOCIATE EDITOR / DESIGNER 助理编辑/设计 » Michael Villardi
DIGITAL CONSULTANT 软件顾问 » Casey Rehm
COVER IMAGE 封面设计 » Michael Stradley
Graham Billings
Elias Darham

RPI FACULTY 伦斯勒教师
COORDINATOR 设计协调 » Gustavo Crembil
DIGITAL ADVISOR 软件指导 » Jefferson Ellinger

TONGJI FACULTY 同济教师
Hong Chen 陈宏
Wei Wei 魏崴
Hongwei Liu 刘宏伟
DIGITAL ADVISOR 软件指导 » Ecrument Gorgul 郭安筑

TRANSLATION 翻译
Jennifer Yifeng Zhao 赵艺枫
Tao Jia 贾涛
Bing Bai 白冰
Peter Sen Zhang 张森

PHOTOGRAPHY 摄影
Gustavo Crembil
Hong Chen 陈宏
Joseph Hines
Matthew Sokol
Seth Hepler

4

Contents
目录

FOREWARD 前言
Evan Douglis 埃文·道格勒斯 6
Zhang Jianlong 张建龙 10

SELECTED ESSAYS 论文精选
BUILDING AND WEAVING » Ralph Ghoche 拉尔夫·高彻
建构与编织------------------ 14
WEAVING AS A RESEARCH TOOL FOR URBAN THINKING » Chen Hong 陈宏
以编织主题作为研究手段的城市设计教学尝试------------ 20
CREATIVITY AND CHALLENGE » Wei Wei 魏崴
创意与挑战--------------- 26
WEAVING URBAN FORMATIONS » Gustavo Crembil 古斯塔夫·克伦贝尔
交互情感构成--------------------- 30

WEAVING SYSTEMS 编织系统
Midterm Projects 中期成果 38

PROJECT PROPOSALS 设计项目
Project Brief and Team Clusters 项目简介和团队集群 104
Cluster Projects 团队成果 116
Studio Culture 设计文化 284

Evan Douglis 埃文 · 道格勒斯 « DEAN, RENSSELAER POLYTECHNIC INSTITUTE 伦斯勒理工学院校长

Foreward by Evan Douglis
前言

In the spirit of countless pioneers throughout history that sought to challenge the accepted practices of their time in favor of an expanded view of the world, the new architect of the 21st century must acquire an equivalent entrepreneurial curiosity to effectively respond to the unique global challenges facing our planet today.

Although the education of the architect still continues in many places around the world to privilege local traditions and regional practices, due to the surge of information technology and the rapid globalization of our cities at the turn of the century, the timeless 'reverence for place' is being usurped from one moment to the next by a much more fluid and complex educational model today that favors blended knowledge and interdisciplinary exchange.

The days of acting unilaterally as architects based upon a myopic view of the world is a bygone era. What is becoming increasingly more evident today is that site-specific regimes of knowledge that have had a legacy of success in our profession in the past are inadequate disciplinary sources today in

历史上，无数的先人们曾为了拓展新的视野而向他们的时代不断提出挑战。同样的，21世纪新一代的建筑师们也必须要以同样的开拓精神来应对今日来自全世界的挑战。

尽管在世界上绝大部分地区对建筑师的培养依然停留在对地域化传统的绝对尊崇上，可是我们应当看到，这种传统的"归属感"的培养正在接受着来自信息爆炸和快速全球化教育模式一轮接一轮的冲击。这种更为流动而复杂的教育模式促使我们必须去交叉学科，融合知识。

今天的建筑师不应只停留在对世界的单向思考上。无数的例子已经证明，以传统的眼光将认知仅局限在具体场地上的做法是远不能及时应对每日我们周围环境、经济和技术急剧的变化的。在面对日益复杂的世界时，建筑师需要以一种更具创造性的手段来将建筑学成功带向新的历史时期。

在这个大背景下，伦斯勒理工学院与同济大学建筑系创立的联合设计教学顺应了发展21世纪新一代特大城市的机遇。本书收录了一系列以上海为背景的课题研究，展示了许多对上海这样特大城市的未来畅想。毫无疑问，上海将在新世纪中充当世界的文

Above, Seth Hepler, Shangahi Pudong Skyline.
上，塞特·和普勒，上海浦东天际线

an era of extreme environmental, economic, and technological change. It's a unique moment in history that demands a far more innovative and collective approach on the part of our architectural community if our profession is going to successfully evolve in relation to an increasingly complex world.

Given this larger context, the unique collaboration established between the schools of architecture at Rensselaer Polytechnic Institute and Tongji University, represents a timely opportunity to address the next generation of mega-cities at the turn of the century. Using the great city of Shanghai as a case-study site, the research in this book posits a series of futurist scenarios predicated on cities like Shanghai leading the world in the next millennium as premier cultural attractors and test

化先锋和新技术的试验场，在建筑与城市设施建设的结合中全面地可持续发展。

在此次联合设计的过程中，来自不同思维、文化和教育背景的学生们在崭新而未知的领域进行有益的探索，追求有革命性影响的新一代建筑，其成果令人叹为观止。通过他们充满创造力的、引人瞩目的作品，我们得以一窥未来建筑的无限可能。

我对于来自两所学校学生们的勤奋工作、创造性设计和富有远见的思想表示钦佩。同时我也要向两校参与教学指导的老师们，即来自同济的魏崴教授与陈宏教授和来自伦斯勒的古斯塔夫·克伦贝尔（Gustavo Crembil）教授表示感谢，他们为此次联合设计的成功做出了卓越贡献。此外，来自同济大学的艾库蒙特·高格（Ecrument Gorgul）和来自伦

beds for new-age engineering; merging buildings and infrastructure together as a holistic sustainable project.

Drawing upon a range of different ideological, cultural and educational backgrounds the students entered into new and unfamiliar territory, as they sought to envision the next-generation of architecture as capable of producing awe-inspiring effects and transformative powers. The work they created in the collaborative studio was authentic, wonderfully speculative and compelling on the highest level as it provides glimpses in to an infinite array of possible futures for architecture.

I would like to commend all the students from their respective schools for their hard work, creative engagement and visionary efforts. I would also like to acknowledge the invaluable contribution made by the leadership team of design instructors that oversaw this important collaboration: Professors Wei Wei and Hong Chen from Tongji University and Gustavo Crembil from Rensselaer Polytechnic Institute. In addition, special mention should go out to the digital instructors Professor Ercument Gorgul from Tongji and Professor Jefferson Ellinger from RPI for their insightful technological support.

I would also like to thank the brilliant leadership at Tongji University; Changfu Wu, Dean; Yiru Huang, Vice Dean; Xiangning Li, Assistant to the Dean; Jianlong Zhang, Deputy Director; and Philip Yuan, Assistant to the Director for their enthusiastic support and generosity throughout the collaboration. We look forward to working together as international partners for many years ahead.

斯勒理工学院的杰弗逊·艾林格（Jefferson Ellinger）经验丰富的数字技术支持也发挥了重要作用。

在此我还要感谢同济大学的吴长福院长、黄一如副院长、李翔宁院长助理、张建龙副系主任和袁烽系主任助理等有关领导与教授对此次联合设计教学工作的配合与支持。希望在未来的教学工作中双方继续保持友好的国际合作关系，再创佳绩。

Left, Seth Hepler, Shangahi Pudong Skyline.
左，塞特·和普勒，上海浦东天际线

Jianlong Zhang 张建龙 « DEPUTY DIRECTOR, CAUP, TONGJI UNIVERSITY 同济大学城规学院系领导

Foreward by Jianlong Zhang
前言

CAUP/TJUU-SoA/RPI Joint Studio has gone through several successful years. This studio is very different from most joint studio programs CAUP has with other institutes. Every year the students from RPI will spend the whole semester in Tongji University taking a series of courses, of which the joint studio project with CAUP senior students is the most important of them all.

In recent years the joint studio celebrated a fruitful semester in urban and architectural design using the *weaving* concept, with the auspices of Gustavo Crembil, Hong Chen and Wei Wei. *Weaving* is a concept, but is also a certain way of thinking when facing various elements associated with the urban landscape and architectural design. Architecture takes the heavy responsibility of extending the memory of history and representing current life. It should inherit historical & spatial form and show the possibility for future evolution. The visualization of *weaving* can be understood as different clues from the environment, the element codes generated being based on space and structure, the balance among the community's interests, as well as the behavior of specific users. The braided method of

同济大学建筑与城规学院和美国伦斯勒理工学院建筑系的高年级联合设计已经合作过多年。与一般的联合教学不一样的是，每次来自RPI的学生都要在同济度过整整一个学期的时间，期间经历一系列的课程学习，其中最重要的便是与同济学生一起完成联合设计课题。

近年来，联合设计课程在古斯塔夫·克伦贝尔（Gustavo Crembil）、陈宏、魏崴等老师的主持下，运用"编织"这一概念，启发、指导学生进行城市设计和建筑设计，成果显著。"编织"是一种概念，也是处理城市环境和建筑空间中各种要素关联的思维方式。建筑既要延续历史的记忆，又要承载当下的生活形态；既要传承经典的空间样式，又要提供未来演变的可能性。可视化的"编织"，可以理解为来自环境的不同线索，也有基于空间和结构所生成的要素编码；有来自于社会社区间的利益平衡，更有具体使用者的行为事件。编织的联结方式是形态生成的法则，转译到建筑中则是建筑空间与形态的基本文法。这种联结是一种耦合，是在尊重各自独立性的同时相互作用，并为形态的生成起催化作用。

weaving is the underlying principle of form generation and the grammar of both space and form in the architectural language. *Weaving* is a kind of coupling, an interaction in the respect of an element's independence and a catalyst for the form generation.

Weaving is also a vivid metaphor between the teaching mode of Tongji and Rensselaer. Different cultures, backgrounds and perspectives merge and collide. This is surly a terrific example of an international joint studio program.

"编织"也是对同济大学与伦斯勒理工学院之间教学模式的形象比喻，不同的文化背景、不同的思维方式在教学中融合并激起火花，是国际化设计教学的成功范例之一。

Left, Gustavo Crembil, Shanghai.
左，古斯塔夫·克伦贝尔，上海

Selected Essays
论文精选

Ralph Ghoche 拉尔夫·高彻 « LECTURER, RENSSELAER POLYTECHNIC INSTITUTE 伦斯勒理工学院讲师

Building and Weaving
建构与编织

At first glance, fabrics and textiles, and the knots and weaves by which they are made, would seem to have little in common with architecture. Buildings, after all, are erected with rigid members that when fastened, hardened or assembled, produce an immobile and largely inflexible frame. Textiles, on the other hand, are necessarily supple and elastic in order to hang with the forces of gravity, wrap around a piece of furniture, or adjust to a body in motion. Beyond this, the divergence in rigidity between weaves and buildings is joined by the very different sense of duration that they both prompt. Traditionally, and as its root word would suggest, architecture has historically connoted permanence; comprised of elements that endure. Textiles have an altogether different intimation and suggest the fleeting, the ephemeral; comprised of elements that succumb to the vagaries of time. Clothing, curtains, drapery, textile wall coverings, mats, and screens last rarely beyond a generation, if beyond a season. This is all to say that textile and architecture seem fundamentally different. It is then with some surprise that, in the last two

各种布料，纺织制品，以及打扣编织的过程，看上去似乎与建筑没有什么关联。当我们把一栋建筑的各个部件加固、硬化、组装起来后，所树立起来的应当是一个不可移动不易变形的坚固整体。而相比之下，纺织的结构则极为柔顺富有弹性，可以依顺重力而下垂，包裹家具或者随时根据人体的运动而调整形状。

除此以外，编织与建筑对"不变性"（Rigidity）的理解也因其持久性的差异而不同。无论是从传统的角度还是从其词根来看，建筑的含义自古以来一直与永恒相连，建筑的材料也因而多为耐久的材料。而编织则正相反，它总是寿命短暂。衣服、窗帘、帷幔、墙纸、席垫和屏风通常在用过一代后就过时了，有时甚至都用不了一季的时间。总而言之，建筑与编织之间似乎完全是两个不同的概念。因而在过去的两百年中，不断有人试图从类似编织这样看似偏远的视角去研究建筑的起源、目的和使用，这样的现象也就值得探讨了。

从很大程度上来说，戈特弗里德·森佩尔（Gotfried Semper）是通过对建筑和编织的词源及考古的研究建立了现代建筑学中最深入而隐喻的理论之一。

centuries, a practice so visually and phenomenally distant from building as weaving might be, would emerge as a recurrent trope to explain architecture's origins, purpose and performance.

It was largely on etymological and archeological parallels between the arts of weaving and architecture that Gottfried Semper would erect what is certainly one of the most penetrating modern architectural metaphors. As is too often neglected, Semper's mid-nineteenth-century theory was not conceived out of whole cloth, but had its basis in the work of the French Romantic architects' counter-establishment interpretation of the origins of architecture, particularly Henri Labrouste's fascination with the figure of the knot as a symbol for unity. These Romantic architects were captivated by the historical transformation of ephemeral ritualistic forms (sacrificial offerings, graffiti, and vegetal garlands) into "petrified" architectonic elements made of durable materials such as stone or bronze. Semper developed this bit of insight into a comprehensive theory that sought to extract from the architectonic motifs of his present day the primitive forms underwriting their historical evolution. Of the four primitive elements cited by Semper, wickerwork (and related forms of weaving) carried the most important spatial consequences. Woven screens and carpets hung upright, Semper explained, were the first spatial enclosures, the first walls.[1]

The analytical possibilities the theory afforded were indeed tremendous. Operating as a historical x-ray of sorts, Semper's theory allowed him to see within the staggered courses of a brick

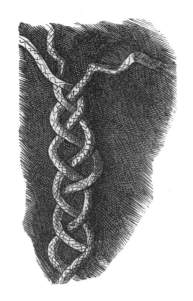

在看待这个问题时有一件事情常常被我们忽略：森佩尔在19世纪中期提出的理论并不是基于"完整的织物"。他的理论是基于法国浪漫主义建筑师们对建筑起源的反传统理解，尤其是亨利·拉布鲁斯特（Henri Labrouste）的以"节扣"作为统一象征的研究。这些浪漫主义的建筑师们深深着迷于历史上那些将稍纵即逝的精美的仪式物固化为石制或是铜制的建筑部件的做法：祭品、涂抹绘画、植被花环，等等。森佩尔将这种看法逐步发展并试图上升为同时代的建筑主题。他的理论所引用的四种基本部件中，柳条编织（以及相关联的其他编织物）是最为重要

1 *Above*, Gottfried Semper, Figure of the Knot, *Der Stil* in den technischen und tektonischen Künsten oder praktische Ästhetik (Munich: Friedr. Bruckmann's Verlag, 1878), 172.

1 上，戈特弗里德·森佩尔，缠结图案

wall, for instance, the faint historical traces of the primitive weave of a screen. The procedure was akin to the work of a biologist (Semper highly admired French natural historian Georges Cuvier) trying to understand the organic structures of living specimens by grasping their transformation from the earliest fossilized remains.

The historical connection that Semper drew between textiles and architecture was a way to rejuvenate the discipline by purging it of two centuries of idealist academicism. In other words, there was a subversive agenda behind Semper's evocative argument that architectural spatiality began with the hanging of woven screens. As a political agitator (Semper had fled from Dresden due to his participation in the May Uprising in 1849), he was conscious that his theory disrupted artistic hierarchies by wresting architecture from the academic fine arts and returning it to the handicrafts. In fact, the entire thrust of Semper's unfinished masterwork *Der Stil* (Style in the Technical and Tectonic Arts) lies in uncovering the tactile and experiential basis behind what would otherwise be seen as flippantly stylistic or abstract aesthetic decisions.

Semper's methodology provides us with a way into the work gathered in this publication, the products of a collaboration between architecture students at Rensselaer Polytechnic Institute and Tongji University. As noted in many of the abstracts prefacing the projects, the rich tradition of craft production in China provided an effective zone of exploration and an opportunity to further entrench the buildings into the cultural imagination of their users. The experiments

的空间秩序制定物。他解释道，编织的屏风和挂毯是最初的空间围合，即是最原始的墙。（图1）

这个理论为建筑打开了无穷的可能性。森佩尔的理论就像一台历史的X光扫描仪，让他能够看穿哪怕是极微弱的历史痕迹。举例来说，他可以将砖墙交错的布置联想到原始的屏风编织所使用的类似手段。正如森佩尔所极为推崇的一名自然历史学家乔治·居维叶（Georges Cuvier）在工作中所做的一样，他通过研究物种从最早的化石开始一系列的变化来研究现有生物有机的结构。

2 *Right*, Louis Kahn & Ann Tyng, Model of a New Office Tower for Philadelphia. Unbuilt (1957).

in woven and knotted webs and meshes that crowd the pages of this book can thus be seen as genuine attempts at rediscovering a kind of indigenous craft with the potential to connect not only the pre-historical with the contemporary, but also the tactile with the digital, the corporeal with the not-yet material.

While the student projects assembled in this publication pick up on some of the ramifications of Semper's theory of weaving, they also depart from others. Chief among these variances is the distinction Semper drew between space and structure. Woven screens and textile walls, in Semper's narrative, were primarily symbolic and spatial enclosures that carried no structural load. Semper's separation of these two elements would have a profound effect on the practice of architecture for well over a century. The idea colored the work of much of the avant-garde, especially that of Mies van der Rohe and Le Corbusier, whose free planning celebrated the freedom of partition walls from the gravitational burdens of the building. Not until the mid-twentieth-century work of Buckminster Fuller, Ann Tyng and Louis Kahn would space and structure be rejoined. The projects here seem to derive their spatial intuition from this latter trajectory. Like Kahn and Tyng's exuberant tetrahedral city tower projected for Philadelphia, the designs assembled here see in the geometrical tissue of the weave the possibilities for unitary space-structure construction. The weave here has become a strategy for responding to the exigent complexities of the program, the users, and of the site.²

森佩尔在编织与建筑之间所建立的联系正是可以用来摆脱两百年来统治建筑领域的理想学院派的理论,振兴建筑学的出路。换句话说,森佩尔所指出的"建筑空间始于手工壁毯"的言论背后有着一种颠覆性的力量。身为一名政治改革家(森佩尔在1849年因曾参与五月起义而被迫逃离德累斯顿Dresden),他清楚地认识到他的理论打破了传统观念上的艺术等级制度,把建筑从美术学院中解放出来,回归到传统手工艺之中。事实上,森佩尔未完成的著作《技术与构造艺术中的风格》(Der Stil Style in the Technical and Tectonic Arts)中全部价值就在于它揭示了编织背后有关触觉和实验性的基础,而并非大众所普遍理解的编织仅是随意的样式或是抽象的美学。

森佩尔的理论将带领着我们走过这本伦斯勒/同济联合设计集。其中很多项目在前言中也提到了,中国极为丰富的传统手工艺文化正适合于学生们对这一课题的探索,也方便了在人群中建立建筑与文化印象的联系。本书中的一系列有关编织和结网的设计项目都可以看作是一系列从新发掘传统手工文化的尝试。它们不仅将历史与现在联系在了一起,而且也把触觉与数字技术,实体与虚拟相连接。

本书中的学生作品是基于森佩尔的学术成果,但也有所不同,尤其是在关于空间与结构的关系的问题上不大一致。在森佩尔的介绍中,编织的屏风和交叠的墙仅存在于象征和空间意义上。它们并非承重构件。这种空间与结构分离的看法在其后的一个世纪中一直有深远的影响。像密斯·凡·德·罗和勒·柯布西耶这样前卫的建筑师十分青睐于去从承重墙中解放出自由隔断墙,并由其去组成自由平面。一直到20世纪中期才出现像巴克敏斯特·福乐

Here too we find a historical thread relevant to our enterprise, for textiles and the computation of complexity have a shared origin in the brilliant programmable loom invented at the dawn of the nineteenth century by French weaver Joseph Marie Jacquard. As many have noted, Jacquard's loom, which automated thousands of different weaves by preparing specialized wooden punch cards for the task, was among the first machines to translate complex real world patterns into abstract computational vectors. The same operative logic is found in the genus of buildings that emerged at the advent of the era of electronic computation as reactions to modernist functional segregation.

These "mat" projects as they were called (a term coined by the Brutalist architect Alison Smithson), sought to exploit the textile analogy by conceiving of buildings as intricate networks producing complex interactions between otherwise distinct functional programs. The "groundscraper" designed and built by Candilis-Josic-Woods for the Free University in Berlin was undoubtedly the most celebrated mat building: the "groundscraper" designed and built by Candilis-Josic-Woods for the Free University in Berlin was undoubtedly the most celebrated mat building of the postwar era.[3] One is reminded of Jacquard's loom but also of Candilis-Josic-Woods innovative webs when looking at the exotic nets and weaves cast in this book. As many influential architect have shown throughout the past two centuries, there is something fantastically exploratory in the way that design projects can translate woven fabric into built form. The

（Buckminster Fuller），安妮·唐（Ann Tyng）和路易斯·康（Louis Kahn）这样希望从新将空间与结构结合的设计师。这里学生们的作品更倾向于后者，在几何形体上把空间与结构统一起来，这种手法在一定程度上与安妮·唐和路易斯·康在费城所设计的城市大厦有几分类似。学生们在书中的编织项目首要解决的是功能、使用者以及场地的复杂性。(图2)

这些"地毯型"设计（受野兽派建筑师艾莉森·史密森（Alison Smithson）的理念而得名）把建筑理解为复杂网络下产生的复杂交互关系，并以此来探索和类比编织的过程。康迪利斯·约赛克·伍兹（Candilis-Josic-Woods）在柏林自由大学设计的"拓展型大楼"（Groundscraper）绝对是战后时期"地毯型"设计的代表作。(图3) 我们可以从本书不同的编织设计中看到约瑟夫织布机以及康迪利斯·约赛克·伍兹的"网络"的影子。和两百年来著名的建筑师们所作的不断尝试一样，今天这本设计集中学生们的作品把编织的形式演化到具体的建造形式上的诸多尝试让我们叹为观止。可以预见，无论在微观尺度上还是宏观尺度上，编织和节扣仍将是未来建筑师们设计复杂建筑系统和结构的试验场。

projects in this publication demonstrate that knots and weaves, here thoughtfully investigated at both micro- and macro-scales, continue to be a rich terrain for architects to produce complex architectural systems and structures.

3 *Above*, George Candilis, Alexis Josic and Shadrach Woods, Aerial photograph of the Free University, Berlin (date unknown)
3 上，康迪利斯·约塞克·伍兹，柏林自由大学鸟瞰（日期不详）

Hong Chen 陈宏 « PROFESSOR, TONGJI UNIVERSITY 同济大学教授

Weaving as a Research Tool for Urban Thinking
以编织主题作为研究手段的城市设计教学尝试

It's so exciting that the students' project of the 2012 Tongji-RPI joint studio will be published as a book. Beginning in 2000, the biennial joint design studio of the two schools has been continuously held uninterrupted seven sessions. Tongji has a history of collaborating with foreign institutions to carry out joint design teaching. I must thank the auspices of RPI professor and coordinator of this joint studio—Gustavo Crembil. It is his meticulously organized efforts and bountiful energy to edit and organize content that ensured this book smooth publication. He whole-heartedly taught in the joint studio as well as showed high sense of responsibility and the spirit of selflessness in the work. Here is my review of the whole joint studio:

Time and Personal Arrangements » The 2012 joint studio was held in February through April as was the case in previous years, which is a time period where architecture senior students have access to the free-topics courses. Configuration of the students put Chinese students and American students at a ratio of 1:1. Both sides of students are free to enroll; two students make one group to start the joint project, the length of project being

2012同济——RPI联合设计学生作业成果结集出版实在是一件令人高兴之事。从2000年开始至今两校间的联合设计每两年一届已连续不间断地进行了7次，在同济与国外多所院校开展联合设计教学的历史中RPI是较早的一个美国院校。两校合作至今已从本科生联合设计、我校青年教师赴RPI进修访问进入到目前正在推进的硕士生、博士生联合培养的全方位合作模式。当然本书得以出版必须感谢RPI此次联合设计主持教师古斯塔夫·克伦贝尔（Gustavo Crembil），正是在他的精心组织下，并花费了大量精力和心血对内容进行编辑和整理从而保证了此书得以顺利出版，一如他在联合设计教学上的全情投入和高度的责任心以及忘我的工作精神。下面对此次联合设计情况作一个回顾：

时间及人员安排 » 2012年的联合设计如往届一样安排在2月～4月举行，与我校建筑学专业四年级学生第二学期的自由选题课相对接，根据美方学生的来访人数按1：1人员配置中方学生，双方学生均自由报名参加，两人为一组开展联合设计，合作设计时间为8.5周，在整个教学过程中，美方师生在同济，双方师生从开始到最终成果展评均全程参与。

8.5 weeks. In the whole process, RPI students and professors stayed in Tongji, both schools' faculties and students were in full participation from the beginning to the final review.

In the joint studio teaching activities of our school with several other universities in the world, like RPI, compliment each school's syllabus content and engage otherwise rare collaborative teaching.

Site and Project Selection » All of the previous Tongji-RPI joint studio projects are situated around the rapidly changing developments in Shanghai. We choose different sites every time, from the reconstruction of Old City to the development of the waterfront area to redesign the urban of the complexity area. In recent sessions, Gustavo Crembil has acted as the program coordinator from RPI. With so many joint studios, each session makes the communication and cooperation between each school more tacit and gradually forms a clear design theme based on site conditions. In the last session, the site delineation was in a large city-wide in the Hongkou district, and the professors encouraged each group of students to choose different specific sites for their project. The design theme is 'Urban Hiatus', which hopes that students can use the theory of urban design to organize the city space, to connect the history and future as well as create passion in city.

For this session, the Shiliupu area was selected as the site of this project. The North edge of the site has been marked by the Shiliupu comprehensive transformation phase I of the project and can be extended to the Bund on foot; the South side facing Fuxing Road where the old pier urban transformation

在同济与全球多个院校开展的各种类型的联合设计教学活动中，像RPI这样在时间与内容上完全对接各自学校的教学课程大纲并进行全程合作教学的模式并不多见。

基地与设计选题 » 同济---RPI的历届合作设计课题均围绕上海这一快速发展与变化的城市展开，每次选择的基地各有不同，从旧区改造、滨水区域发展到城市复杂地段空间重整等等。RPI的近两届院长Alan Balfour和Evan Douglis均对上海这座城市表现出了浓厚的兴趣，认为这是一座充满历史和活力的城市。Alan Balfour并曾经在美国出版了《上海》的英文专著。

近几届RPI来访的负责教师均为古斯塔夫·克伦贝尔（Gustavo Crembil），双方教师的多次合作使得相互间的沟通与合作更为默契，并逐步形成根据基地条件和环境明确设计主题并开展设计的教学思路。

上一届的基地选址是由教师在虹口区划定一个大的城市范围并与各组学生共同讨论选择其中不同基地进行研究，设计主题是"都市之隙"（Urban Hiatus），希望学生运用城市设计方法去缝合和组织城市空间，连接过去与未来，创造城市活力。

本届的基地选在了上海南外滩十六铺沿黄浦江区域，地块北面是已建成的十六铺综合改造一期工程并可步行延伸至外滩，南面临复兴路接老码头城市改造地块，西面临城市交通主干道中山南路，东面则是黄浦江。基地呈狭长条状，面积约4.5公顷。基地内还有20世纪90年代建成建筑面积有3.4万平方米的交通银行大厦高层建筑，另外城市地铁九号线在基地下面穿江而过，西北方向离地块不远就是豫园。此地块是外滩沿江区域最后一个待开发的用地，它

plot is; the West side facing the main traffic line—Zhongshan south Road; the East side is Huangpu River. The site is a long and narrow strip, the area being about 4.5 hectares. In this site, there is the Bank of Communications Tower, which is a 1990s era building that has an area is 34,000 square meters. In addition that, the Metro Line 9 runs through underneath the site, and Yuyuan is not far away from the northwest side. This site is the last undeveloped area in the Bund waterfront area. The urban design should be effective to drive the project, to create the urban form and organize some public activities to stimulate the vitality of the entire area in this city. The design theme is 'A Great New World'.

In order to meet the teaching requirements of the weaving joint studio, the students were only required to meet the function of urban planning of Shiliupu Dock Phase II, other functions such as office space, businesses, culture, residential and hotel were all freely configurable. There were no limits for planning indicators such as floor area ratio and coverage ratio.

Organization of Teaching » The development of the discipline of contemporary architecture, in the context of globalization, is no longer a static value system, as the diversity of the people's spiritual needs also bring multiple values. In today's society, the technical and economic aspects that had protected built-in, traditionally Western architectural concepts are no longer the only means of evaluation of building systems. Architecture has become more impacted by a variety of disciplines such as sociology, psychology, biology and philosophy. In such a social context, the openness

的城市设计定位应该是能有效带动周边已开发和正在开发项目，延伸外滩的城市公共活动人流并连接到老码头和豫园，创造城市形态，组织公共活动，激发整个区域的都市活力。设计主题是"新的大世界"（A new Great World）

为适应联合设计编织主题教学需要，对于基地内的功能安排除要求学生需满足十六铺二期客运码头的城市规划功能以外，其他如办公、商业、娱乐、文化、居住、酒店等功能都可根据自己的调研和方案发展自由配置相应功能，并对诸如容积率、覆盖率等规划指标不做限制。

教学组织与形式 » 目前基于全球化背景的当代建筑学科发展已不再是一成不变的价值体系，人们对精神需求的多样性也带来了多元的价值观，当今社会的技术和经济水平也保障了各种建筑构思都有了建成的可能性，传统意义上的美的均衡已不再是建筑的唯一评价体系。建筑学也越来越多地受到社会学、心理学、生物学、哲学等多种学科的影响。在这样一种社会背景下，联合设计教学内容的开放性和实验性就显得尤为重要。

本届联合设计教学的主题是"编织"。

关于在设计教学中引入编织这一教学形式在上一届的教学过程中已经有了初步尝试，本届设计开始前，美方教师克伦贝尔已预先作了充分的教案准备工作，设计开始后同济方面也安排了教师Ercument做了关于编织的专题系列讲课。

从基本概念来讲编织是由一组线通过经纬方向的运动、交织、再运动生成面乃至空间体系。

教学希望让学生先经过这样一种在经纬度及运动方向上的编织技巧训练，去体会各种元素在线形起伏

of the teaching of the joint design and experimental component are particularly important The theme of this joint design teaching is *weaving*.

The introduction of *weaving* in the teaching was attempted in the last collaboration. Gustavo Crembil fully prepared the teaching before the start of this session. Professor Ercument Gorgul was also appointed by TongJi University to deliver a lecture series about weaving projects since the studio started.

Basically, *weaving* is composed of a set of lines that, through movement in latitudinal and longitudinal direction, mixing, and further movement on higher levels of complexity, generate surfaces and spatial systems. With such training in latitudinal and longitudinal motion of weaving, students are expected to experience the relevance and possibilities in the rolling development of elements of all kinds. From there, the relationship between the site and the city, and the different needs within the site will be analyzed and woven together. With motion, deformation, dynamic shape and wavy structure in skin textures, a new site experience is created and a customized architecture-urban relationship can be formed.

The *weaving* theme in the teaching process can be divided into three stages with reviews and grading by the end of each one. Each stage is worth 30 points, with 10 extra points awarded for the creativity and initiative of the student during the whole design process.

Basic Training Of 'Woven' Theory » The teaching organizational arrangements: three weeks before their base analysis, students engaged in literature

发展时的关联性和联系的多样可能性,并获得多种视觉形态的体验。然后以这样一种思维方法去分析并编织基地与城市、基地内多种需求因素的关系。它的主要形态表象有运动和变形、动感的形体和表层肌理起伏的结构等,最终获得一种全新的场地体验经验和个性化的建筑与城市形态。

编织主题在整个教学过程中的安排可分为三个阶段,三个阶段结束时均进行评图和打分,每个阶段各占30分,另外10分是学生在设计全过程中所表现出的学习自主性和创造性。

编织基本训练 » 在教学组织中安排了前面三周时间做基地分析,相关文献阅读和编织基本训练,学生通过电脑设计编织图案并制作单一线型材料的手工编织模型,编织的训练过程中学生基于线的逻辑组织、面的生成、空间生成,需独立完成三个统一尺寸的线、面、体电脑和手工模型。

Above, Gustavo Crembil, Photograph of South Bund Site.
上,古斯塔夫·克伦贝尔,南外滩基地照片

reading and knitting basic training, students through computer design researched knitting patterns and created a single linear model of hand-woven materials.

In this stage of weaving training, both sides of students enjoyed the fun of labor, and put great enthusiasm and high-intensity into their work. The phase of the evaluation criteria is woven logic, growth in change and creativity.

Construct City Relationship » This is the middle three weeks of stage design teaching, students return to the site, using weaving analysis and organization to apply to the city on three levels:

Restrictive conditions (i.e. orientation, flood control, urban traffic, the existing building, etc.)

Flexibility factors (i.e. the border, moving line, form, theme, etc.)

Recessive conditions (i.e. culture, behavior, mood, etc.)

Each group of two students at this stage entered by combing into groups of 4, exchanging different cultural backgrounds & thinking collision is the main theme of this stage students learning at this stage of research and analysis and generate a concept, the students needed to open their minds. The outcome of this stage is the analysis diagrams and abstract entity model, evaluation criteria of the city of woven relations, the concept generated being logical and reflective of personality.

'Woven' architecture » This is the late stage of joint design, the cooperation of the students are in-depth. We have to deepened the concept design

该阶段编织训练对双方学生来讲均具有极强的新鲜感和手工创造的乐趣，学生以高强度的工作状态投入了巨大的热情，该阶段的评价标准是编织的逻辑性、生长与变化、创造性。

建构城市关系 » 这是设计教学的中间三周阶段，要求学生回到基地，用编织的思维去分析和组织城市三个层面上的线索：

限制性条件（如朝向、防洪、城市交通、已有建筑等）

灵活性因素（如边界、动线、形态、主题等）

隐性的条件（如文化、行为、意境等）

学生在该阶段进入两人一小组、四人一大组的合作设计阶段，交流和不同文化背景的思维碰撞是这一阶段的主旋律，学生在这一阶段的学习主要是研究和分析并生成概念设计，思路需要开放，对部分习惯了具象操作的中方学生来讲会有茫然和不适应感。该阶段的成果是分析图和抽象实体模型，评价标准是城市关系编织和分析的合理性，概念生成的逻辑性与个性体现。

编织建筑 » 这是联合设计的后期阶段，学生们的合作进一步深入，要针对设计主题"新的大世界"将概念设计进行深化，调整落实功能配置，梳理动线和公共空间的连贯性与相互作用，编织形式语言并创造多元化城市形态，然后要运用建筑性能评价软件对建筑群体的风和光环境进行测试，并优化和调整建筑形态，最后制作完成展板并配合成果模型。最终评图环节双方院长均会参与讲评和讨论。

编织教学的意义 » 从编织角度研究和开展城市设计教学，以编织作为思维训练的工具，这和我们以

for the 'A new Great World' design theme, adjusted functional configuration and the weaving language to create a complex urban form; using performance software to test the wind and light environment of the building group. Lastly, the students need to finish the boards and the final model.

The meaning of 'Woven' teaching » To carry urban design teaching, we used weaving as the tool of design thinking. This is a big difference to our previous teaching, beneficial for developing the student's mind, which reflects the openness and experimental teaching purposes.

High-intensity learning and teaching » After each class, the RPI professor will send mail to students about a large number of design tasks and data that need to be completed. Very high-intensity work ethic was characteristic of the students in the entire learning process, but only through rigorous training can students grasp the depth of teaching content.

Concern of strategy and form generation » Different subjects design depth requirements, different learning angles for students. Shorter design times allow students to complete the theme as clear and complete as possible, not seeking the perfect design results, but focusing more on students' thinking training. The trend of building in the context of globalization is a concern; studying the theoretical basis of the morphological appearance in this context is the purpose of the weaving themes as teaching activities. Of course, it is only a preliminary attempt. We need to improve and continuous development during the future of teaching and learning.

往的教学有很大不同，对学生设计思路的开阔会带来帮助，也体现了开放性和实验性的教学目的。

高强度的学习和严格教学 » 每次课后美方教师均会给学生发邮件布置下次上课需要完成的大量设计任务和阅读资料，学生在整个学习过程中强度很高，只有通过严格训练的学生才能深入掌握教学内容。

关注设计策略的提出和形态的逻辑生成 » 不同的课题设计深度要求不同，学生学习的角度也不同，对于此次基地的复杂城市环境、较短的设计时间，让学生完成主题明确完整的概念设计训练和表达。不求设计成果的面面俱到，而更注重学生思维的方法训练。

关注全球化背景下世界建筑的发展潮流，并研究其形态表象下的理论基础，是此次以编织作为主题开展教学活动的目的，当然这只是一次初步的尝试。并有待我们在今后的教学中去不断发展和完善。同济——RPI联合设计作为一个长期的教学合作项目给我方的教学活动增加了国际视野，丰富了教学形式，也将会不断推动教师思考如何在全球化背景下进一步提高我们的教学水平。

Wei Wei 魏崴 « PROFESSOR, TONGJI UNIVERSITY 同济大学教授

Creativity and Challenge
创意与挑战

The partnership between CAUP of Tongji University and RPI SoA has gone through more than ten years since the undergraduate joint studio started in the beginning of this century. The biennial joint studio for teacher-student interaction greatly contributed to the development of both institutions. With excellent professors and students from each of the two colleges, as well as different cultural backgrounds, the cooperative education in the research of urban design in the field of developing Shanghai is one of the main focuses in architectural education at both institutes. From the earlier projects like the improvement of the riverside environment of the Suzhou River, the protection and development of the historical blocks on Shaoxing Road, to the Hongkou urban block integrating design practice. In observation, cognition, participation and practice, the TJU-RPI joint studio program is moving forward steadily.

While the Shanghai urban development enters into a mature stage with more organized urban planning, the rise of the financial center in Pudong-Lujiazui and the great success of the 2010 Shanghai World

同济大学建筑与城市规划学院与美国伦斯勒理工学院自本世纪初第一次联合设计教学战略合作至今已走过十余年。两年一度的联合师生交流学习，大大地促进了双方院校的发展。基于双方院校良好的师生资源、不同的城市生活、文化背景，以上海正在发展中的城市为设计研究课题的联合教学项目已成为双方建筑教学的焦点。从早期的上海苏州河沿岸的城市环境改良项目，到上海绍兴路街区历史文化的保护与发展研究，以及上海虹口区地块集合设计，由观察到认知、参与到实践，TJ-RPI联合设计教学工作逐步深入并且持续发展。

今天上海城市发展日趋成熟，城市设计的步骤也更加有序。浦东陆家嘴金融中心的崛起、2010上海世博会园区的建成，标志着上海正从地域建设的"苏州河时期"迈向国际化发展的"黄浦江时代"。

2012年度TJ-RPI 联合城市设计教学为了紧密结合上海城市发展需求，选择十六铺二期地块作为区域城市设计的研究主题，侧重结合参数化设计技术手段，通过教学探讨，研究未来城市多义性和多元性建设发展的可能途径。课题以城市、街区、建筑三个基本

Expo, the development of Shanghai is transforming from the local development of the 'Suzhou River time' to an international 'Huangpu River era.'

The TJU-RPI joint studio in 2012 meets the requirement of city development in Shanghai in choosing the Shiliupu project Phase II as the research topic. With the emphasis on parametric design and academic discussion, our research concentrated on the possible path toward the diversity of the city development in the future. The research developed around the topics of city programmatic areas, transportation and integrated environment design within the setting of the three basic elements: city, neighborhoods, architecture. The weaving skills practice, the simulation of digital technology, physical model studies and other steps highlight the creativity and challenges of the study. It is hoped that through this research, an open city space can be formed.

Parametric Design / Space Syntax » Parametrics has entered the field of design in recent years. It is widely used and continuously developed. In the field of architectural design and urban planning, it is used for the research of complex relationship between architecture and the city's function and morphology via the language of mathematics and logic. Through the adjustment of factors applied into the functional variables, and with the help of powerful computers, different forms of urban model can be simulated, evaluated and compared. This is also one of the main concepts of parametric design in an architectural training application.

In 1970, Professor Bill Hiller of the University of London proposed the Space Syntax theory, as a

要素的设定、实物编织技巧学习、计算机软件模拟、物理模型研究等多个步骤，综合展开对城市功能定位、交通关系、周边环境整合的关联性研究，生成开放的城市空间形态，设计研究过程充满创造性并极具挑战。

参数化设计/空间句法: 参数化设计近年来进入设计领域并广为使用、不断发展。就建筑设计及城市设计而言，则是运用数学语言和逻辑研究建筑和城市的功能与形态之间的复杂关系，并通过设定城市与建筑要素转化为函数变量的不断调整，强大的计算机运算，生成不同的形态关系模型，并最终物化为可评估、可比选的城市空间模型。这也是将参数化设计概念引入建筑设计训练的目的之一。

1970年伦敦大学的Bill Hiller教授提出了空间句法的理论，将其作为一种新的描述建筑与城市空间模式的语言。其基本思想是对空间进行尺度划分和空间分割，分析其复杂的关系，核心内容是在城市的物质空间内对建筑空间和社会关系量化地描述。空间句法也可以作用为建筑师对参数化设计以及城市现状的理解和认知。

择取对象: 建筑综合体/区域的城市形态。关于城市形态的设计研究通常是基于城市的功能定位、交通组织、心理行为等一系列要素的综合量化关系而展开，相对抽象而宏观。研究的重点是确定编织要素间的联系方式、相对位置、空间尺度、形态生成等复杂性问题。而建筑空间创作的目标则是将这个空间中的要素通过构件、联络路径和逻辑结构的有机组合获得心理的、连续的视觉上的愉悦。

学生在计算机模型和物理模型的研究尝试将从一个新的视角去观察、理解、发现区域城市空间形态对

description of a new mode of architecture and urban spatial language. The basic idea is to subdivide the scale and space and to analyze the complex relationships among them. The core content is how to cognitively develop architectural space and the social relations in a materialized urban space. Space Syntax can also be understood as a way for architects to perceive the parametric design process as well as the status of the city.

Choose the object: The urban form of building complex/area » Urban design is usually based on the abstract and macro environmental characteristics such as programmatic function, traffic organization and the psychological behavior of people. The focus of the study is to determine complex issues like the manner of nodes, spatial position and scale, form generation and so on. The goal in the creation of architectural space is to achieve psychological and continuously visual pleasure through the organic combination of the elements, linked paths and logical structure within a range of spaces.

Students are expected to observe, understand and discover the effect of environmental performance and public behavior driven by urban space from a whole new angle with the help of computer software and physical model studies. The Phase II of Shiliupu project is significant for the urban development process, not only in terms of its typical urban landscape, but also in its complex transportation and the positioning of its building program. Being able to solve this problem on the many differing levels of the urban environment and its elements, and to coordinate the various programmatic

周边环境以及人的行为活动所产生的综合影响力。十六铺二期地块在上海城市发展进程中具有一定的代表意义，既包含了城市层面的景观要素的典型性，也兼具区域地块的交通的复杂性以及建筑功能定位的综合性。解决城市环境的不同层次的问题、协调各种功能空间组织的可能性，形成未来城市的创意形态，正是本次课题学习研究的动力所在。

企及目标：实验操作/教学思考。通过计算机模拟与编织物理模型的两种手段的并用，教学工作从空间环境、区域城市形态、建筑综合建构三个方面选定

Right, Matthew Sokol, Photograph of South Bund site.
右，马修·索科尔,南外滩基地照片

functions together, which may creatively generate future urban form proposals, is the motivation of this research.

Target: Experimental Operation/Teaching reflection » With the means of computer simulation and the knitting physical model studies, the research focuses on the complexity and variability of the generation of architectural form. The spatial environment, regional urban morphology and building integrated construction marks the three key levels to set the parameters. Facing specific conditions with an abstract experimental method, accompanying the macro urban positioning with the micro architectural variables, evolving the spatial potential to develop multi-directionally, the outcomes of the whole process is full of uncertainty, but indeed it is the great effort from all the teachers and students who contribute their intelligence and creativity. As a factor of the comprehensive study of urban design, the research attempt to break through the assumption of a pure rational thinking and the current technical standard, also puts into question the possible ways or methods of survival for the city of the future.

With new technology on the design approach and a ground-breaking thinking on the issue of urban form, the research meets the educational principle of Tongji University - 'Inclusive Harmony', and the motto of RPI 'Why not change the world?'. I'm looking forward to seeing the creation and discovery in next the season of TJU-RPI joint studio.

关键参数，重点研究建筑形态生成的复杂性和多变性。以抽象的实验方法面对具体的环境对象，宏观的城市区位定性分析辅以微观的建筑要素变量，衍生出多方向发展的潜力空间。尽管实验教学的结果尚存在不确定性，但操作过程充满了师生合作的智慧和创意。作为城市设计问题综合研究的一个重要因子，尝试突破单纯理性思维的假设，逾越现行建造技术规范的束缚，同时也质疑了未来城市生存的途径。

在设计方法上对新技术的运用，对城市形态问题的突破性思考，既吻合了同济大学"兼容并蓄"的办学理念，也符合RPI的校训："Why not change the world"。谨以此期待下一季TJ-RPI联合设计教学的再创造、再发现。

Gustavo Crembil 古斯塔夫·克伦贝尔 « PROFESSOR, RENSSELAER POLYTECHNIC INSTITUTE 伦斯勒理工学院教授

Weaving Urban Formations
交互情感构成

Hiatus[1], definition:

1a: a break in or as if in a material object: a gap.

b: a gap or passage in an anatomical part or organ.

2a: an interruption in time or continuity: a break; especially: a period when something (as a program or activity) is suspended or interrupted.

b: the occurrence of two vowel sounds without pause or intervening consonantal sound'.

In the laboratory of urbanism that is today's Chinese city, an architectural hiatus appears between the coarse-grained urban vision of centralized planning and the tangible reality of building and development. As defined by the oppositional praxis and experiences of the planned (the idealized 'power point city') and the organic (the everyday city), there is fertile ground for new programmatic speculations and architectural procedures in this chaotic and unexpected gap.

裂缝[1]，词典释义：

1a: 在物体之上或者仿佛在物体之上的裂口；近义词：间隙。

b: （解剖学）组织或器官之间的空当或通道。

2a: 在时间或者连续性上的阻断；近义词：休止。尤指项目或者活动被暂停或者打断的一段时间。例句：在停笔五年之后。

b: 两个连续的元音字符，中间没有辅音字符。

在对今日中国城市化发展进程的探索中，一种建筑学上的裂缝正在高度集中的城市规划与现实具体中的城市建设之间扩张。源自实践和经验的理想型城市规划与每日城市的有机生长之间的高度对立，促使我们寻求一条出路以解决这个混沌而难以预测的城市裂缝。

学生们的设计被设定在上海黄浦江西岸的南外滩。这里的滨水区不仅是一个重要的上海城区，它对于整个上海市的全球发展战略亦是极为重要的一笔。这一地区有大量集中的城市基础设施，因而就如同上海的缩影，包含了当今所有的发展问题：高架、隧道、码头，保存状况不一的老房子、公园、水滨景观、崭新的商业办公区、待开发的街区，等等。如果说上

1 *Hiatus*, Merriam-Webster Dictionary.
1 裂缝，韦氏词典

Below, Marek Kolodziejczyk. Wool-thread model to compute optimised detour path networks. Institute for Lightweight Structures, Stuttgart, 1991
下，马立克科沃杰伊奇克，羊毛线程模型计算优化的迂回路径，轻型结构研究所，斯图加特，1991

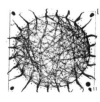
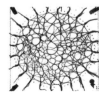

Student projects were to be sited in Shanghai main riverfront, at Huangpu River's West Bank in an area known as The South Bund, an important location that responds not only to local conditions but also largely to Shanghai's global scale skyline imaginary. This is an area with intense urban and infrastructural characteristics. It is a microcosm where all the pressing issues of contemporary Shanghai are brought to the edge: highways, tunnels, docks, old housing in different stages of conservation or dilapidation, parks, scenic promenade and waterfront, brand new towers and commercial districts, blocks (empty or demolition ready) awaiting development, and so on. If Shanghai is the quintessential Chinese cosmopolis, both ruthless and sophisticated, then this urban fragment shows a kaleidoscopic panorama of its urban dramas and dreams.

The studio aimed to address the tectonic argument by scaling the textile notion to the urban realm. Students were asked to develop affective strategies that weave programmatic, environmental and structural forces into an urban scale architecture. Structural systems of textiles were researched and explored by both physical and computational means in order to understand their embedded intelligence and techniques.

According to Ali Rahim, '*affects*, therefore *affective formations* are generated through techniques'.[2] In this studio, the interworking techniques (knitting, weaving) were the exploratory tools used in the research. There is a dialectical polemic between the notions of effects and affects – effects

Above, Travis Lydon. Urban study diagram, 2012

上，特拉维斯·林顿，城市研究图表，2012

海具有典型的中国式国际大都市那种既高强度又细致的发展特点的话，那么这种碎片式的城市布局正展现了一种城市发展中独特的千变万化的景观。

此次联合设计尝试将纺织技艺应用于城市尺度设计的手法。学生们将项目、环境与组织结构融合于城市尺度的建筑设计之中。在实物与数字手段的结合的学习过程中，学生们探索研究了纺织的技术精髓。

正如阿里·拉西姆（Ali Rahim）所解释的，"相互影响的生成关系是通过技艺实现的。"[2]。在此次联合设计中，交互的技艺（如针织，纺织）均作为研究中探索性的工具。这里需要解释直接影响（effects）与交互影响（affects）之间的论证关系：直接影响是单向的影响，是出自当初的预期的结果。而交互影响是主体与客体之间双向影响，双方均可以去影响也可以被影响。我们希望通过对编织技艺的模仿而生

2 Ali Rahim, 2006.

Right, Dana Shin. Weaving Study, 2012, plastic
右, Dana Shin. 编织研究, 2012, 塑料.

3 Ali Rahim, 2006.
3 阿里 拉西姆, 2006

4 Irene Emery, 2009, p. 27
4 艾伦 艾美瑞, 2009, p.27

understood as the result of one-way causality (or intended outcome); and affects understood as two-way interaction between subjects and/or contexts (the possibility to affect and be affected). Moving away from textile analogies that shroud conventional building forms, we were interested in the development of novel forms and spaces, and in the 'production of *affective formations* – works of architecture that maximize their affects and hence their responsiveness to users and contexts'.³

In the majority [of fabrics], separate *elements* (each with its own *structural make-up*) are systematically interworked to form a coherent material. The particular nature and order of the *interworking* is what distinguishes one such fabric structure from another, and consequently classification of the structures rests on classification of the *systems of interworking*. For this it is necessary, first, to determine the number of *elements* or *sets of elements* on which each system is based.

Element:

    a component part or unit of the structure of an interworked fabric. The term refers to yarn, thread, strand, cord, thong, or whatever natural or contrived unit of fibers or filaments is interworked to form a fabric.

Set of elements:

    a group of such components all used in a like manner, that is, functionally undifferentiated and trending in the same direction. Whenever certain elements are differentiated from others

成建筑形态，并在此基础上去探索一种全新的建筑形式。这种新形式将能最大化地与用户以及城市背景交互影响。³

在织物的结构中，相互分离的个体（每个个体具有独立的结构）被系统的组织起来以形成一个连贯的整体性材质。个体之间自然属性的差异以及组织方法的差异是区别任意两种纺织结构的根本。进一步说，对于结构的定义是基于对编制过程的方式的定义。因此，我们的首要任务就是确定个体的数量，或者是个体的编组，并以此为基础建立系统。

个体：

编织结构的组成部分或单元。这个词适用于纱、线、链、带，以及任何可以编织在一起的自然或人造的纤维状物体。

个体的编组：

一组同等处理的个体，在功能上还是走向上都是相同的。无论是改变走向还是改变结构功能，当其中一些个体被区分而独立出来的时候，它们就形成了新的编组。⁴

in the same fabric, either in the direction they take or by the purpose they serve in the structure, they constitute a separate set of elements.[4]

Students were asked to develop a series of physical and digital studies of basic types of 'fabric structures'. Using Irene Emery's fundamental catalog as reference, they developed series of three-dimensional tessellations (vertical and horizontal) by interworking elements through linking, looping, knotting, and their more complex combinations (interloping, interlinking, linking and looping stitches, etc.). These conceptual series were to be deployed in the site, where localized conditions were reworked within the 'original fabric' into a multi-programmatic structure (retail, culture, hotel, transportation) that would situate itself within the historic city fabric and other new developments.

Craft as Haptic Thinking » Aside from identifying and interpreting constraints and opportunities to be used as conceptual groundwork for design proposals, these explorations aimed to instantiate manufacturing and fabrication as scales pertinent to the field of architectural design. This led to the development of basic proposals that were able to weave space and tectonics into comprehensive pieces of architecture. These works can be seen as forthright but humble attempts to intersect, without mediation, oppositional ways of thinking and making, with an understanding that both are traversed with a similar ethics of craft.

学生们通过发展一系列的实体与数字模型来研究基本的纤维结构。借助艾伦·艾美瑞（Irene Emery）的分类研究成果，学生们通过各种编织手法建立了一系列三维的细面划分（包含垂直方向与水平方向）。连接、环接、扣接，以及所衍生出来的更为复杂的编织关系。这些概念性的设计将被布置于基地之上，并与现实的环境状况相结合，建立一个包含零售、文化、酒店、交通等项目的多功能结构来与城市已有的，以及待发展的区域编织在一起。

在工艺中体验触觉 » 基于对编织概念设计中的限制与机遇的认识，学生们的这些探索将建设与编织具体结合到了建筑设计之中。正如最初所设想的将空间与构造组织到具体的建筑实例中来一样，这些学生的尝试直率而谨慎，打破固有的对工艺的认识，把理论与实践结合在一起。

Below, Joseph Hines and Cui Muhan. A New Great World, student project, 2012.

We must put aside the notion that craft is defined by a reliance on tradition, primitive methods, handiwork, a high degree of affect, or a feudal division of labor. The essence of craftwork lies in its acceptance of variation, its incremental pursuit of mastery or virtuosity, its reliance upon embodied or tacit knowledge, and its acceptance of the fundamentally open-source dimension of work. Craft is adaptive behavior applied to non-routine, attentive, collective tasks of fabrication. Craft is haptic thinking.[5]

This studio aimed to contributed to our schools' studio culture and continue to actively broaden the discussion between digital and analogical thinking and making. We are at a critical moment where the way we relate to and represent the world is radically changing. How are we to sensibly act in the world?

I would like to thank the leadership of both institutions, in particular Prof. Yiru Huang and Prof. Jianlong Zhang from CAUP/TJ, and Dean Evan Douglis from SoA /RPI for their trust and continuous support in the development of this program. Finally I would like to thank my colleagues Chen Hong and Wei Wei for their academic generosity and warm friendship, you all made me feel at home in Shanghai.

我们必须放弃这样一种认识：工艺绝对依赖于传统，只得使用传统的方式，必须纯双手制作，完全反映古代的工作分配制度。工艺的真正魅力在于对不同变化的接受能力，在于对更加精湛的技艺的追求，在于对其中隐晦知识的重视，在于其绝对开放的工作方式。工艺是一种非常规的、要求细致而工作量巨大的生产过程。对工艺的思考正是一种依赖触觉的思考。[5]

这次联合设计希望通过双方学院的交流进一步深入对信息化和类比化的思考与制作的讨论。联系与表达我们周围世界的方式正在迅速变化着。如何在这个关键时刻理智地应对这种变化，正是摆在我们面前的课题。

5   Gustavo Crembil and Lynch, 2009, p. 49
5   古斯塔夫·克伦贝尔和林奇, 2009, p.49

35

Weaving Systems
编织系统
Midterm Projects 中期成果

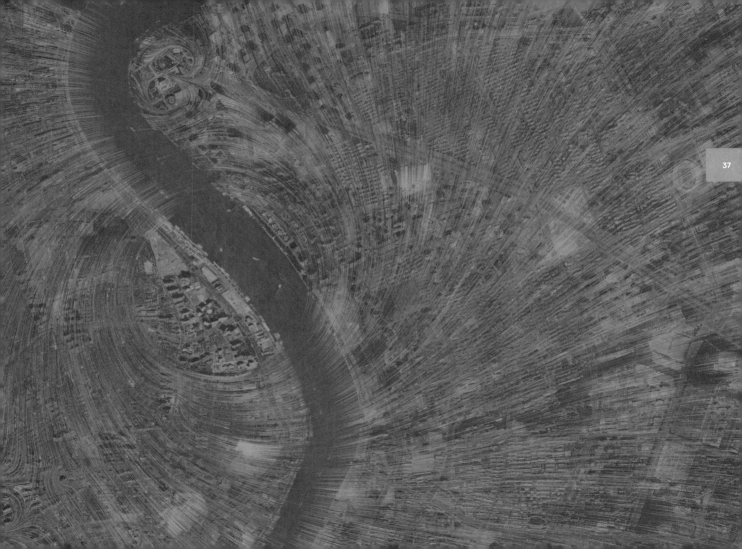

37

WEAVING SYSTEM 编织系统
Danjing Zhao « CAUP TONGJI

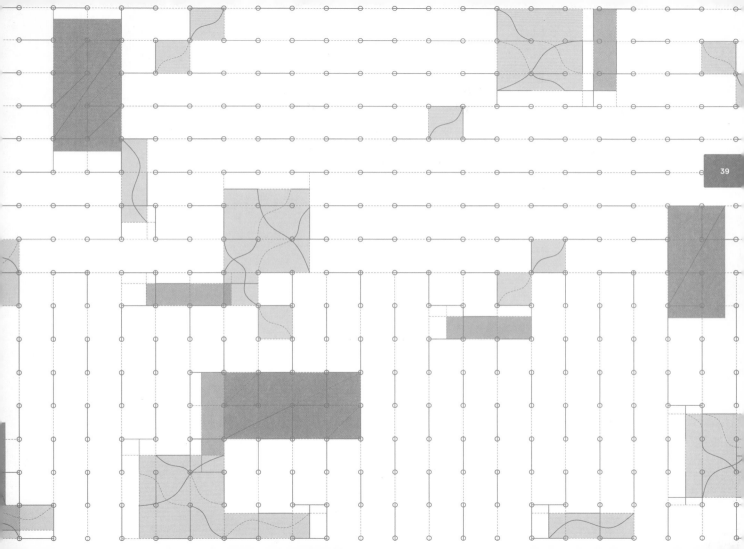

WEAVING SYSTEM 编织系统
Matthew Sokol « SoA RPI

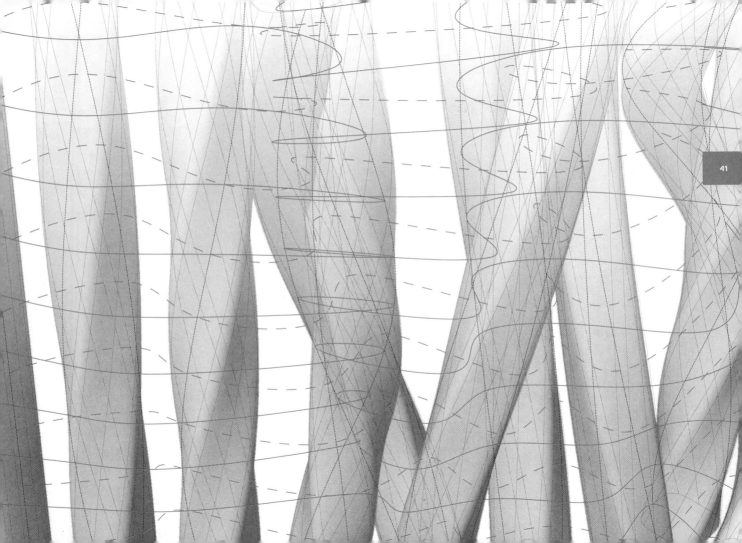

WEAVING SYSTEM 编织系统
Chi Zhang « CAUP TONGJI

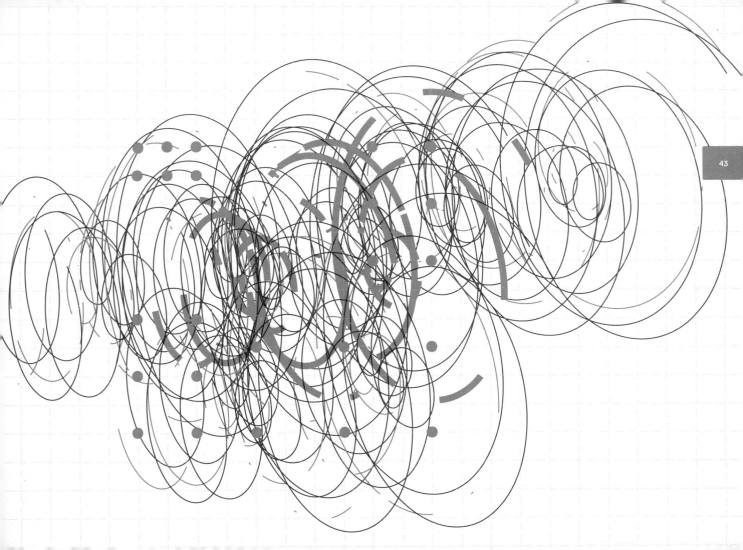

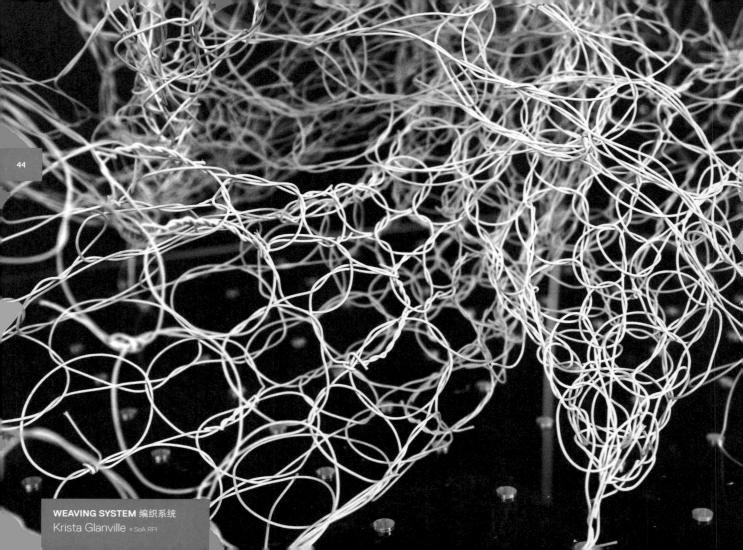

WEAVING SYSTEM 编织系统
Krista Glanville « SoA RPI

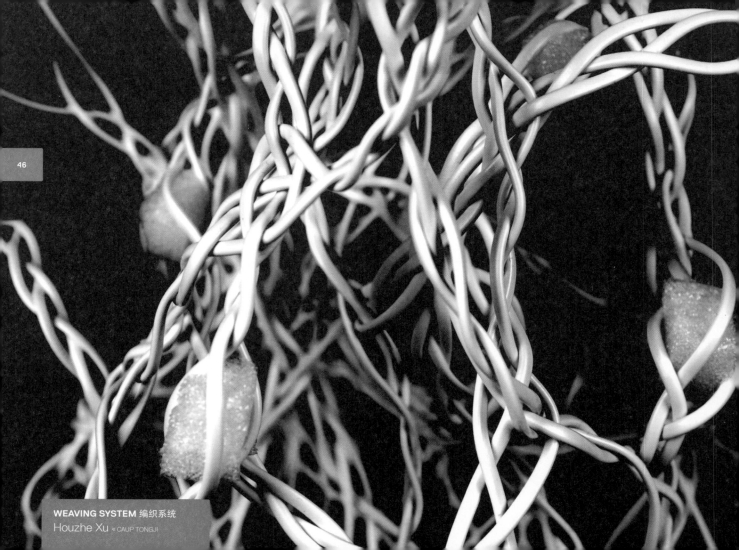

WEAVING SYSTEM 编织系统
Houzhe Xu « CAUP TONGJI

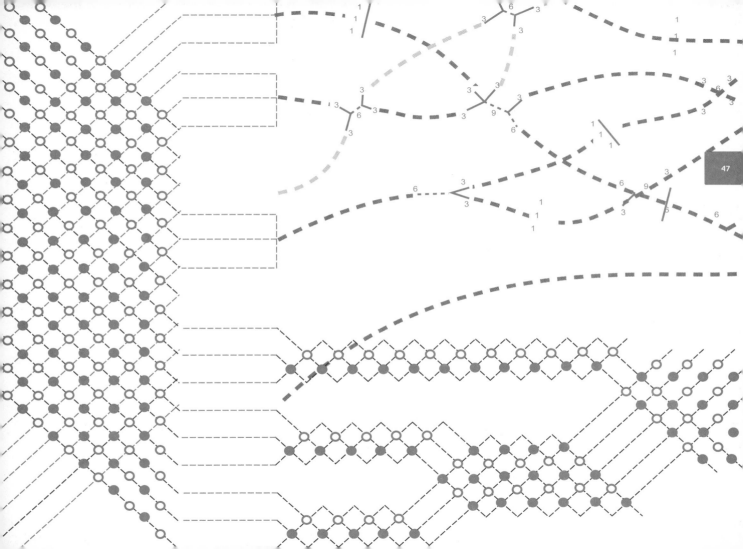

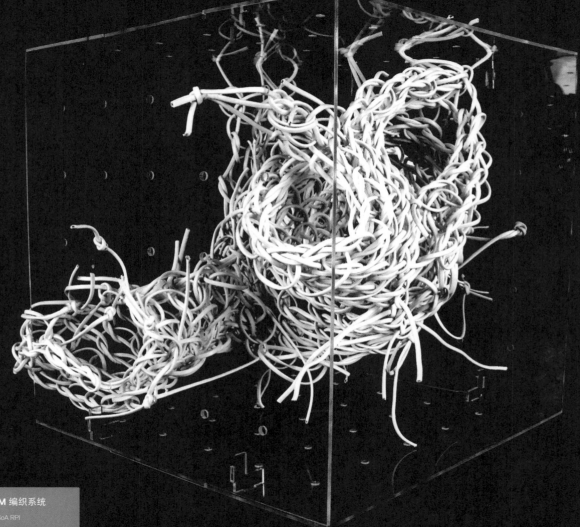

WEAVING SYSTEM 编织系统
Desireé Edge « SoA RPI

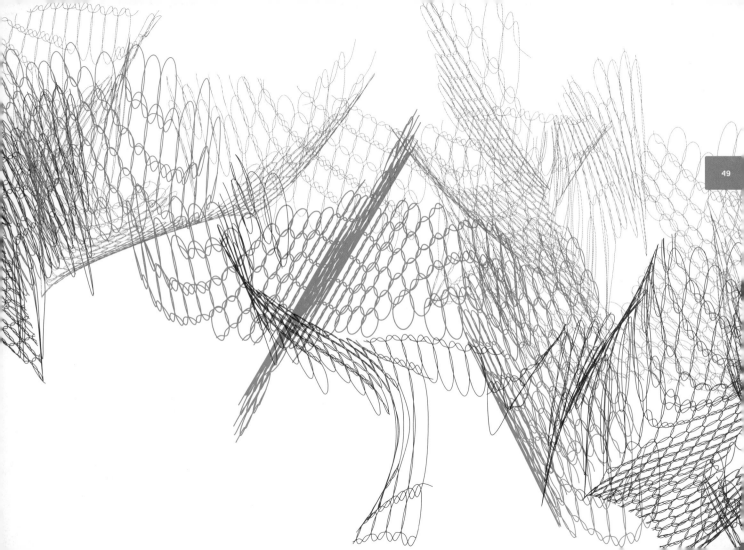

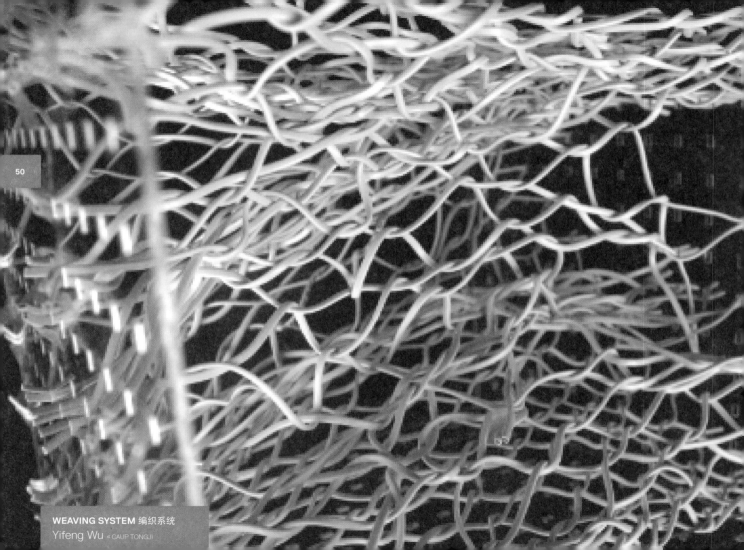

WEAVING SYSTEM 编织系统
Yifeng Wu « CAUP TONGJI

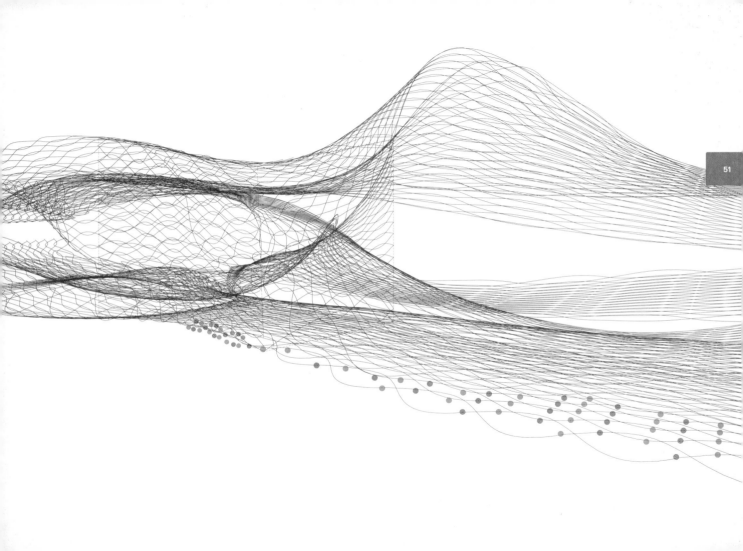

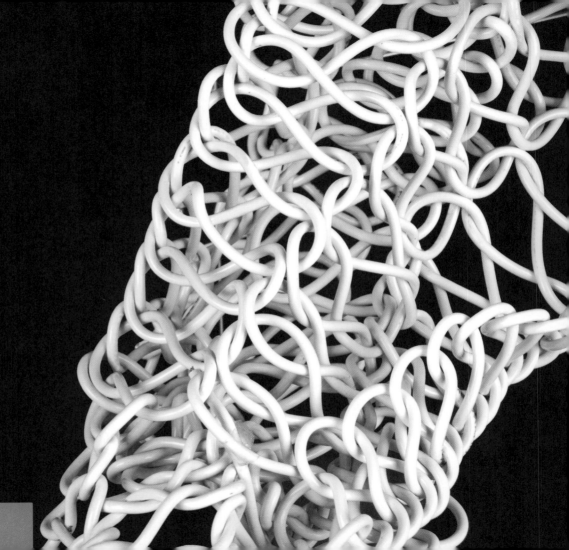

WEAVING SYSTEM 编织系统
Michael Stradley « SoA RPI

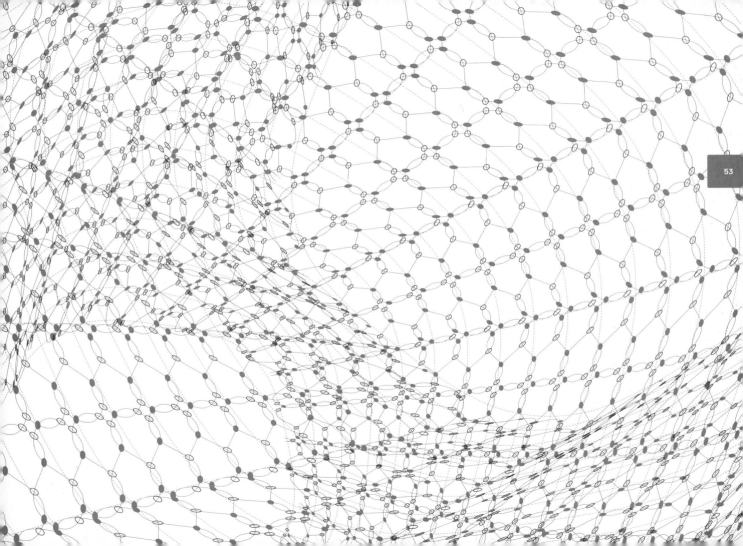

WEAVING SYSTEM 编织系统
Jun Su « CAUP TONGJI

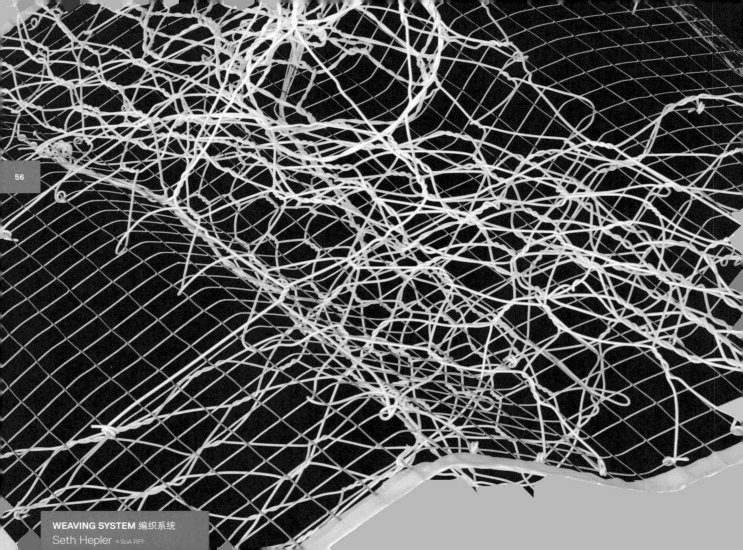

WEAVING SYSTEM 编织系统
Seth Hepler « SoA RPI

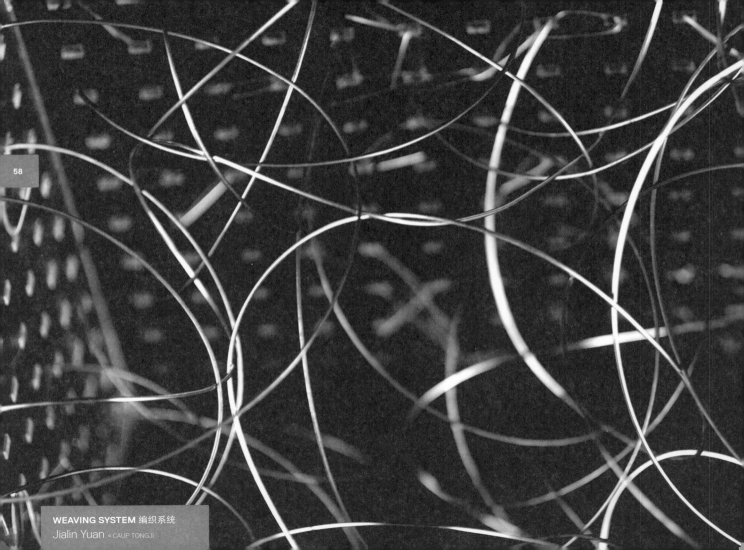

WEAVING SYSTEM 编织系统
Jialin Yuan « CAUP TONGJI

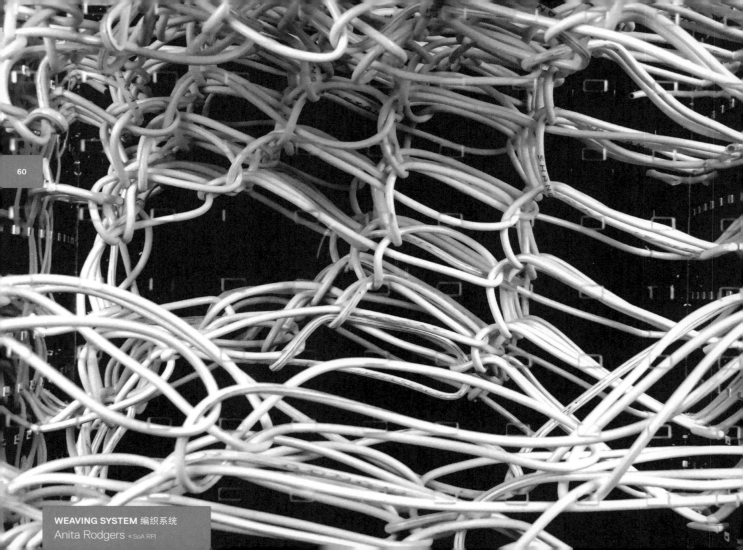

WEAVING SYSTEM 编织系统
Anita Rodgers « SoA RPI

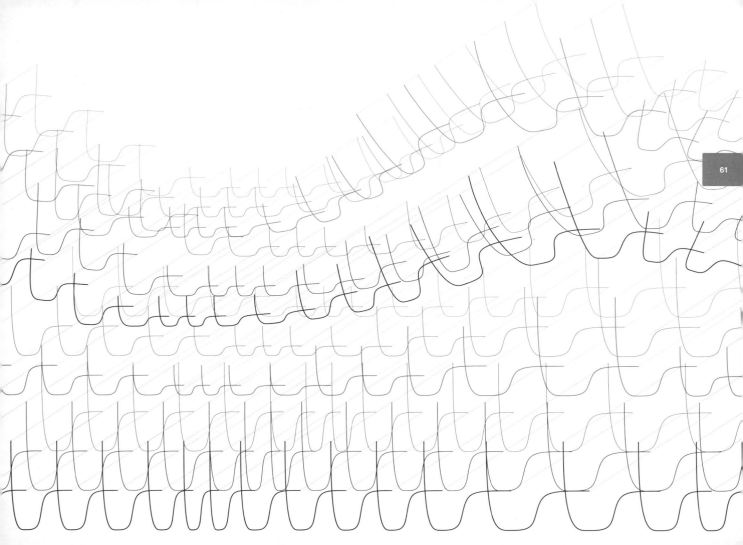

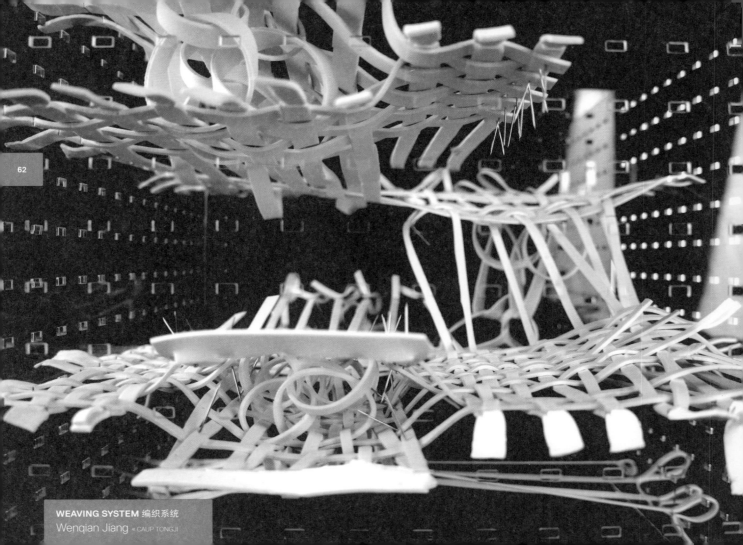

WEAVING SYSTEM 编织系统
Wenqian Jiang «CAUP TONGJI

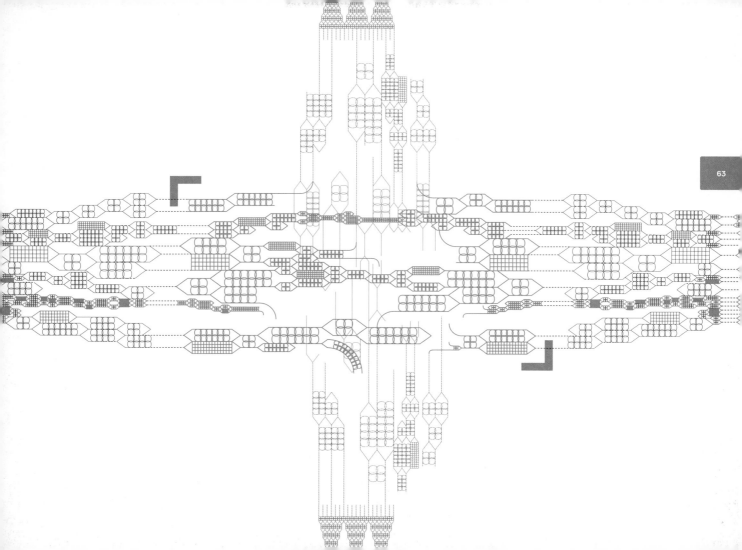

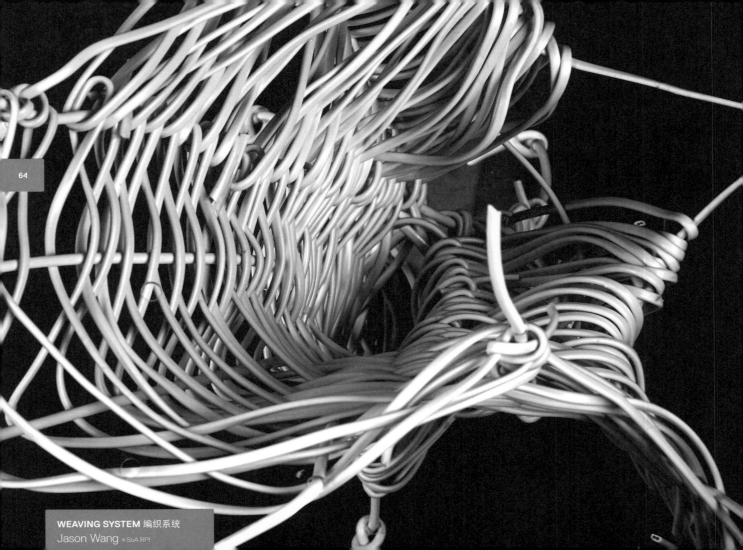

WEAVING SYSTEM 编织系统
Jason Wang « SoA RPI

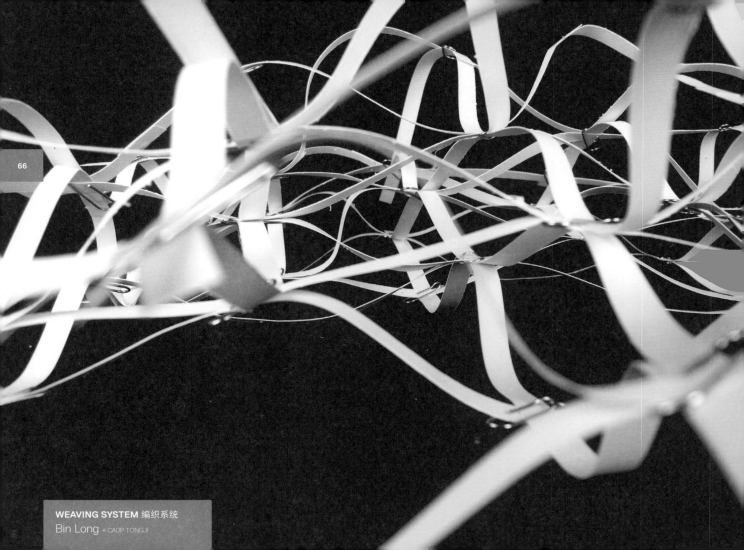

WEAVING SYSTEM 编织系统
Bin Long « CAUP TONGJI

67

WEAVING SYSTEM 编织系统
Travis Lydon « SoA RPI

WEAVING SYSTEM 编织系统
Muhan Cui «CAUP TONGJI

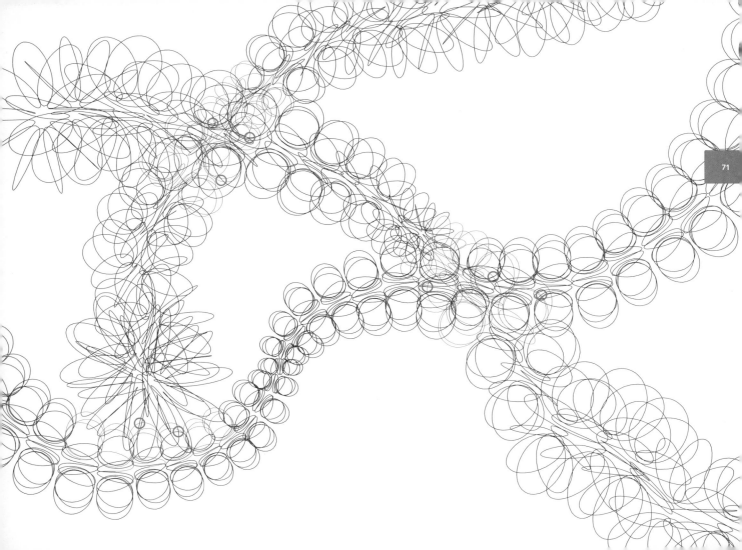

WEAVING SYSTEM 编织系统
Joseph Hines « SoA RPI

73

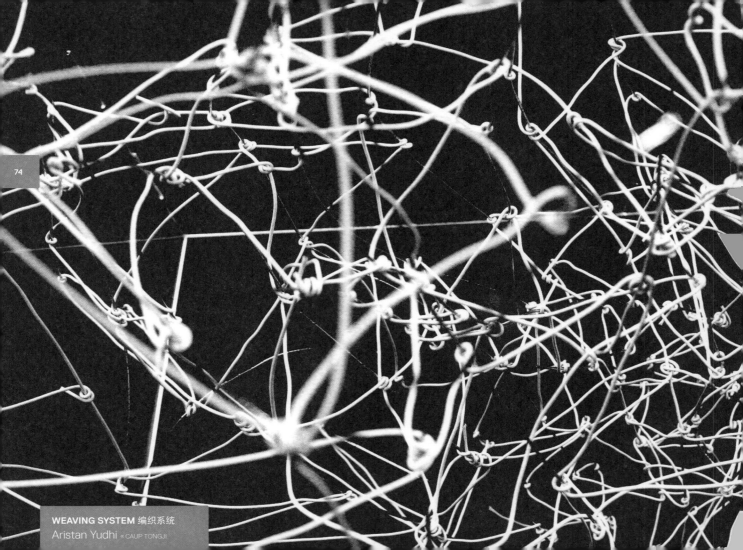

WEAVING SYSTEM 编织系统
Aristan Yudhi « CAUP TONGJI

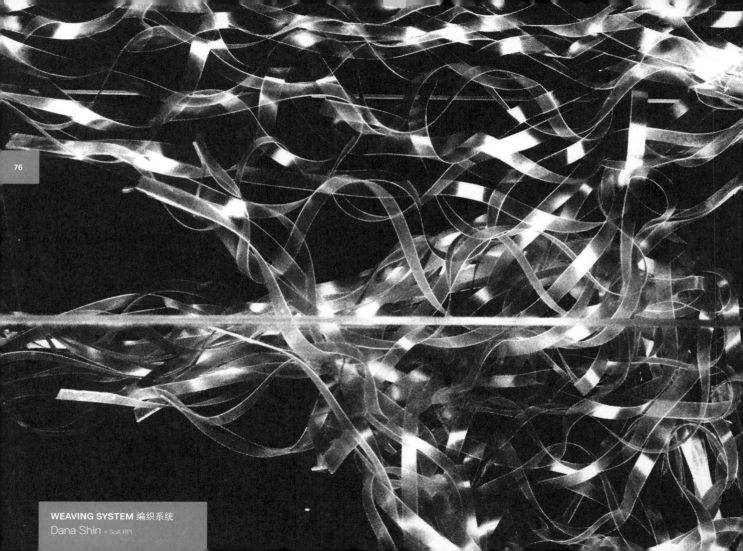

WEAVING SYSTEM 编织系统
Dana Shin « SoA RPI

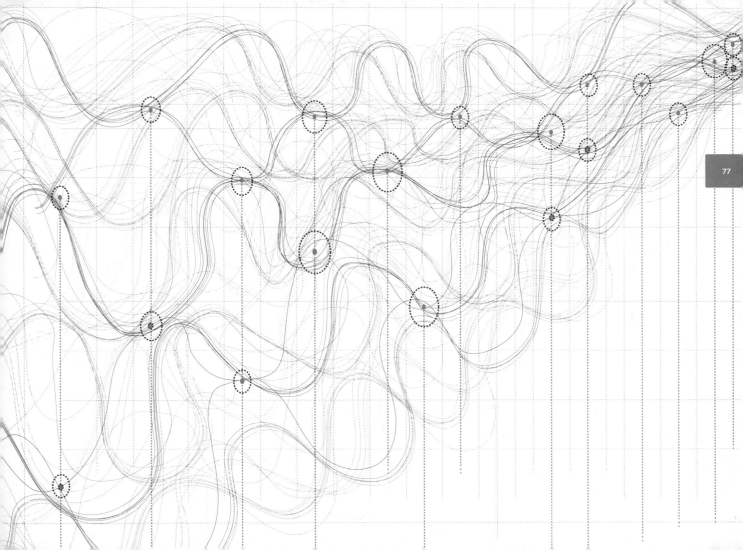

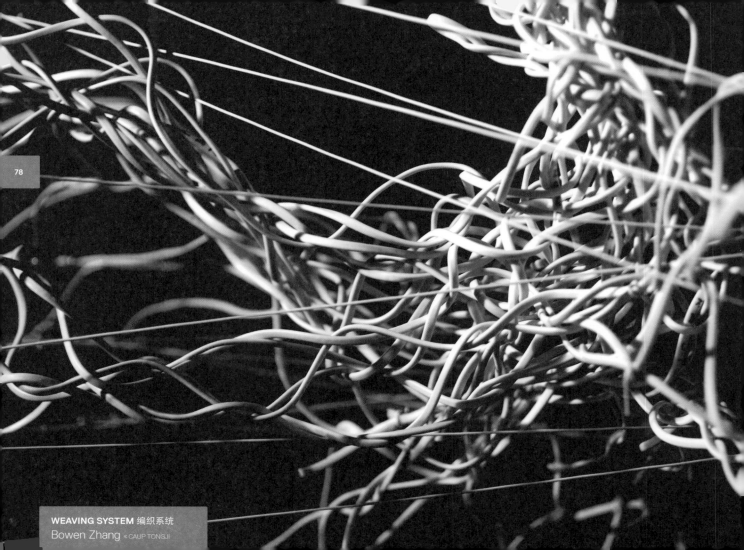

WEAVING SYSTEM 编织系统
Jamie Lee « SoA RPI

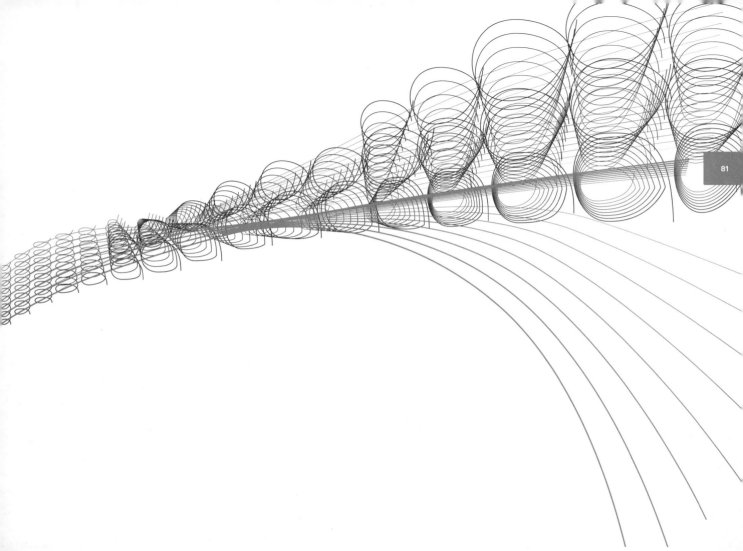

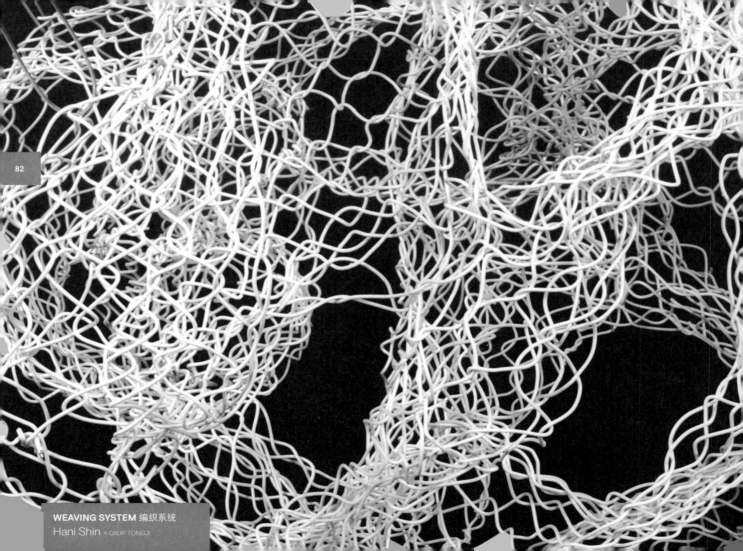

WEAVING SYSTEM 编织系统
Hani Shin « CAUP TONGJI

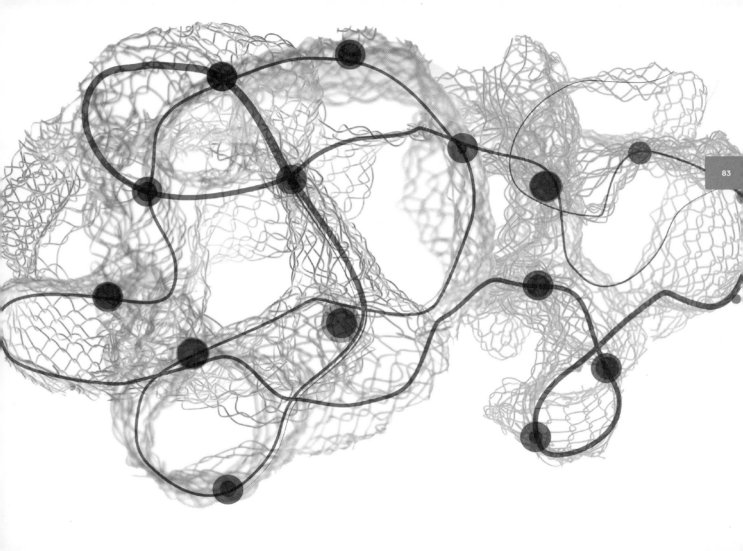

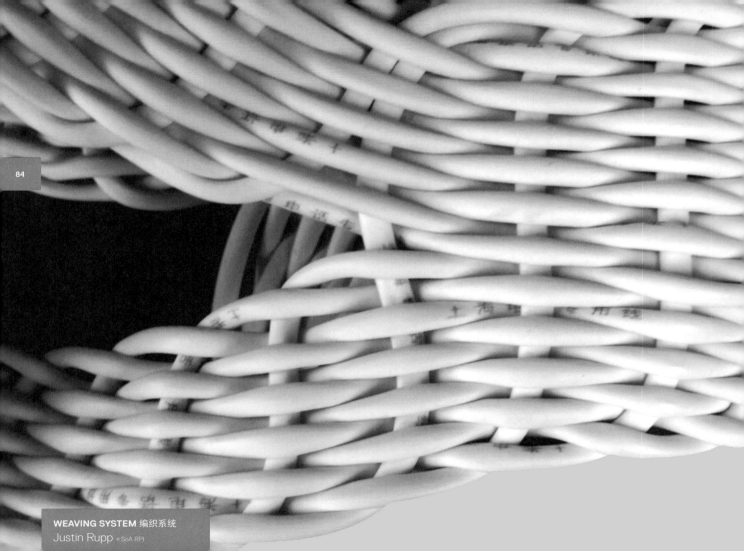

WEAVING SYSTEM 编织系统
Justin Rupp « SoA RPI

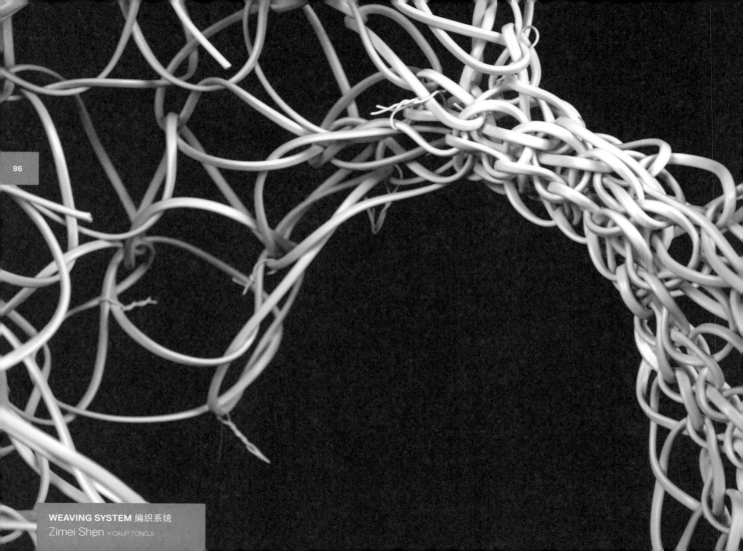

WEAVING SYSTEM 编织系统
Zimei Shen « CAUP TONGJI

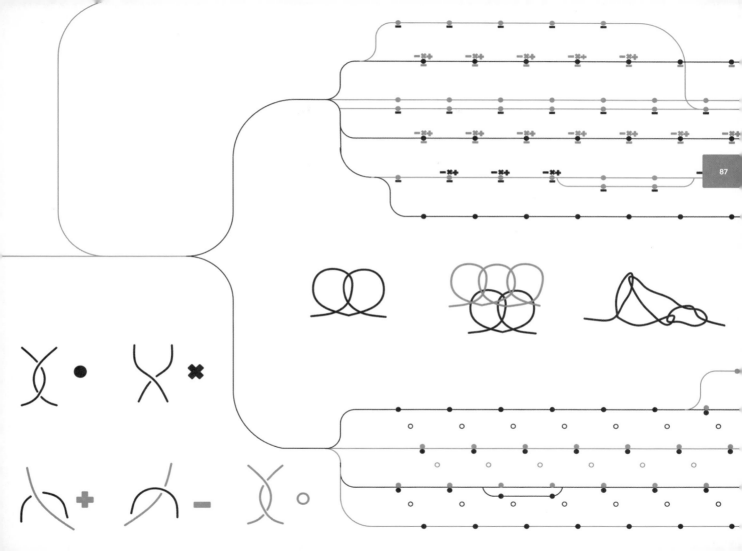

WEAVING SYSTEM 编织系统
Anthony Policastro « SoA RPI

WEAVING SYSTEM 编织系统
Yinjia Gong « CAUP TONGJI

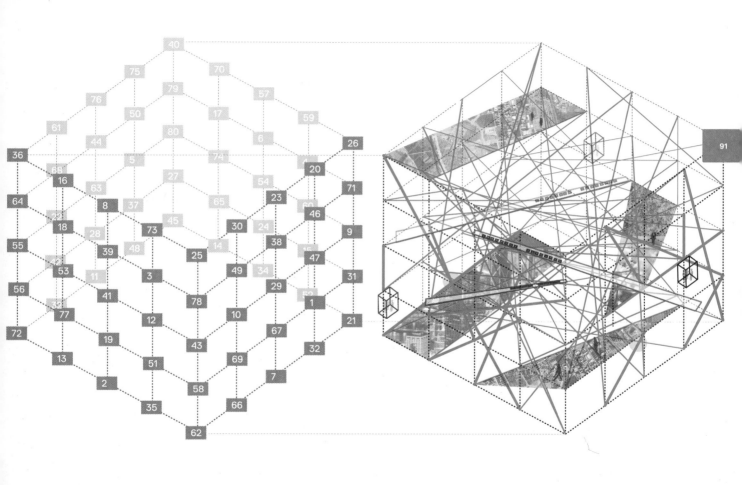

WEAVING SYSTEM 编织系统
Michael Mancuso « SoA RPI

93

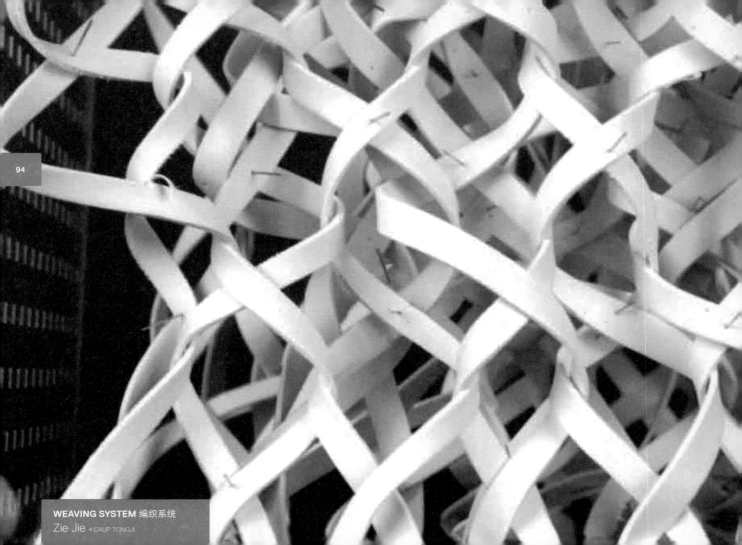

WEAVING SYSTEM 编织系统
Zie Jie « CAUP TONGJI

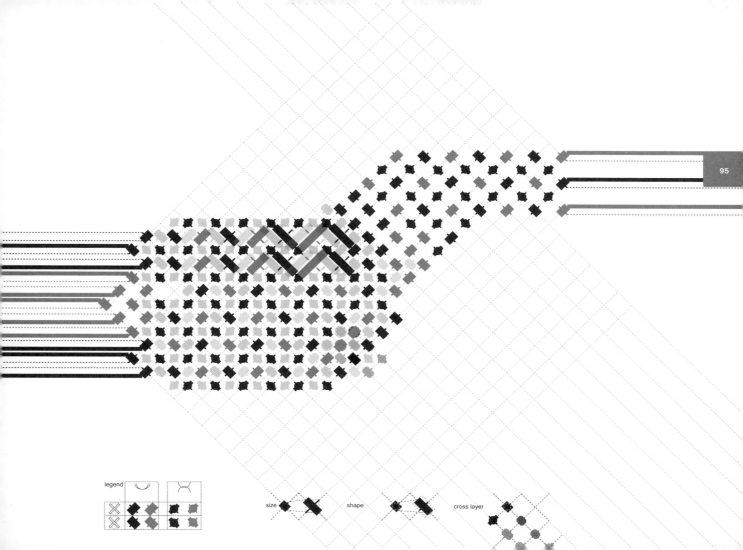

WEAVING SYSTEM 编织系统
Caitlin McCabe « SoA RPI

97

WEAVING SYSTEM 编织系统
Ning Wang « CAUP TONGJI

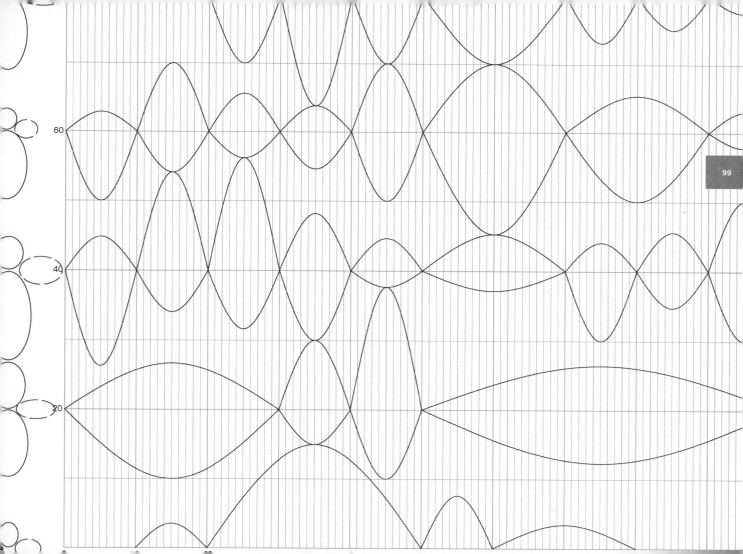

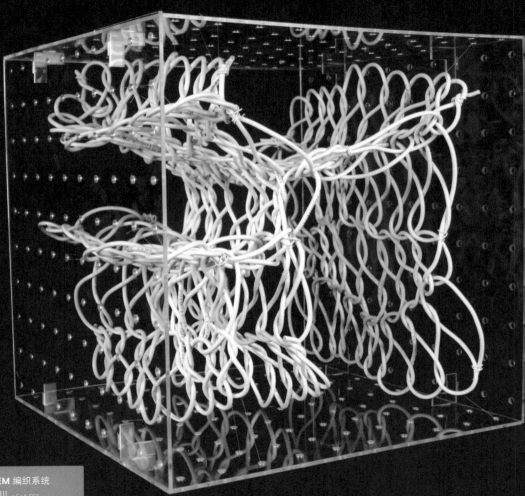

WEAVING SYSTEM 编织系统
Arthur Adams III « SoA.RPI

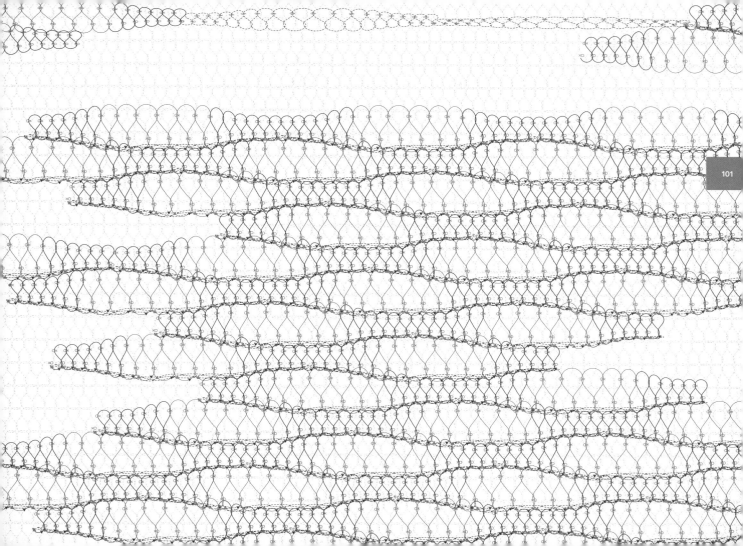

Project Proposals
设计项目

103

Project Brief and Team Clusters
项目简介和团队集群

Above, South Bund area of Shanghai; site of the studio project.

The South Bund » The site is located at Huangpu River's West Bank, in the area known as South Bund where Touristic Ferry and Passenger Terminals currently are. It is roughly rectangular site defined by the Indigo Hotel (North), the river (East), Fuxing Rd (South) and Zhongshan Nanlu (West). It could be understood as continuation of the Bund's promenade towards the South, leading to the new creative industries area of The Cool Docks, facing Pudong's riverside residential developments, and isolated from the Old Town by the a strong vehicular artery and blocks of new commercial and residential developments.

This is an area of intense urban and infrastructural characteristics. A microcosm where all the pressing issues of today's Shanghai are brought to the edge. If Shanghai is the quintessential Chinese cosmopolis, both ruthless and sophisticated, then this urban fragment shows a kaleidoscopic panorama of its urban dramas and dreams.

Great New World » The Great World (Dashijie) was a famous entertainment centre located in the corner of Yanan Lu and Nanjing Lu in the Northeast corner of today's People Square. A multimedia complex before they were invented. The reference of such programmatic vitality will be the precedent for the students to develop a new multi-programmatic structure that articulates itself with the historical Bund, the Old City remnants, and the new developments within the South Bund and beyond. This new program will consider the existing marina and could incorporate a mix of required office spaces, retail spaces, cultural and entertainment facilities, high-end residential spaces, a hotel, and related support programs such as circulation, transportation, and parking.

南外滩 » 基地位于黄浦江西岸，在著名的南外滩和十六铺旅游码头区域，该码头二期项目在基地内，基地形状接近矩形，北临英迪格酒店、东临黄浦江、南临复兴路、西接中山南路。它是外滩观光步行路径往南的延伸，并通过基地连接南面老码头创意园区，临河面对浦东滨水住宅区，豫园与基地相隔一个正在建设中的城市综合体街区并被中山南路交通主干道所隔离。

这是一个有着强烈都市及基础设施特征的地区，今天的上海所有紧迫问题被带到边缘的一个缩影。如果上海是典型的中国大都会，兼具无情和复杂性，那么这个城市的片段显示了一个万花筒般的全景城市戏剧和梦想。

大世界 » 大世界是位于上海延安路和西藏路转角即现在的人民广场东南角的著名娱乐活动中心。一个在多媒体还未发明前的多媒体建筑综合体，同学将参考这样一个内容丰富、富有活力的先例来清晰地表达一个考虑到位于外滩历史风貌区域，有旧城尚存老建筑和南外滩未来发展的新的多元项目的结构发展。这个新的项目将要考虑已经存在的码头以及未来办公、商业零售、文化和娱乐设施、高层住宅、一个酒店及其与之相关联的动线、交通和停车。

Cluster 1 一组
page 118
- **A** 上 — 赵丹晨 / Matthew Sokol
- **B** 下 — 章驰 / Krista Glanville

Cluster 2 二组
page 138
- **A** 上 — 徐厚哲 / Desiree Edge
- **B** 下 — 吴熠丰 / Michael Stradley

Cluster 3 三组
page 162
- **A** 上 — 苏俊 / Seth Hepler
- **B** 下 — 袁佳林 / Anita Rodgers

Cluster 4 四组
page 182
- **A** 上 — 蒋文茜 / Jason Wang
- **B** 下 — 龙彬 / Travis Lydon

Cluster 5 五组
page 204
- **A** 上 — 崔沐晗 / Joseph Hines
- **B** 下 — 陈勇辉 / Dana Shin

Cluster 6 六组
page 228
- **A** 上 — 张博文 / Jamie Lee
- **B** 下 — 新叶丹 / Justin Rupp

Cluster 7 七组
page 248
- **A** 上 — 沈子美 / Anthony Policastro
- **B** 下 — 龚音嘉 / Michael Mancuso

Cluster 8 八组
page 266
- **A** 上 — 谢杰 / Caitlin McCabe
- **B** 下 — 王宁 / Arthur Adams III

113

1ᵃ

CLUSTER PROJECTS 团队成果

STUDENTS 学生
Matthew Sokol «SoA RPI
Danjing Zhao «CAUP Tongji

Towers of Progress
进程塔

High-rises are dominating new developments in cities, but along with the high density solution of this form, a high-rise is fundamentally flawed and in turn can become a hindrance to the city it was meant to develop. A fundamental analysis of the ground condition as a means of transportation, commercial activity and dining revealed an important social aspect which begins to be lost in typical towers. This analysis extends to the surrounding of the site and Shanghai's rapidly receding old city. The aim was to transform such a ground condition into the towers, creating social areas where people would be able to gather and interact. A public park at the 135M level allows the people to enjoy serenity as they look down at the city. The Huangpu River, a vital artery for Shanghai's logistics, is brought into the site. As it passes under a tower the river is brought into the architecture and enables passing boats to engage with the built condition. The building itself serves as a medium for connection resulting in reassuring unity of ground, water, steel and sky.

在现代城市的建设过程中高层建筑是必然的趋势。但是在带来高密度解决方案的同时，高层建筑也有本质上的缺陷，甚至有可能成为城市发展的障碍。像地面交通、商业活动以及就餐等极为重要的社会行为活动正在典型的塔楼建设中消失。通过分析老城厢和场地周边环境，本方案希望在高层建筑中重现这样的城市地面社区，创造一个人们可以在其中聚集交流的公共空间。在135m的高度上设置的空中花园可以让人们欣赏到整个城市的景观。上海的动脉黄浦江被引入基地，引入的河流将从塔楼的底下流过，身在江中的船只也能够进入建筑环境。建筑本身此时已成为一个媒介，联系着大地、江水、钢铁和天空。

Towers of Progress

Axonometric Diagram » A triangulated systems indicating the various plan layouts at critical points in the project.

轴测图示 » 一组三角形状的系统显示在项目关键点的多样平面设计

↗ **Primer Lines**
基本线条

↗ **Critical Points**
关键点

↗ **Lines Between Points**   关键点间的连线

↗ **Finished Form**   已完成的形态

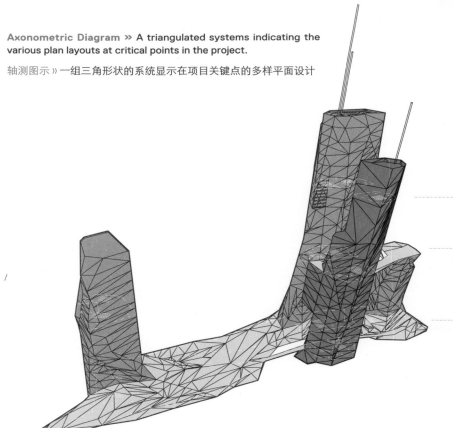

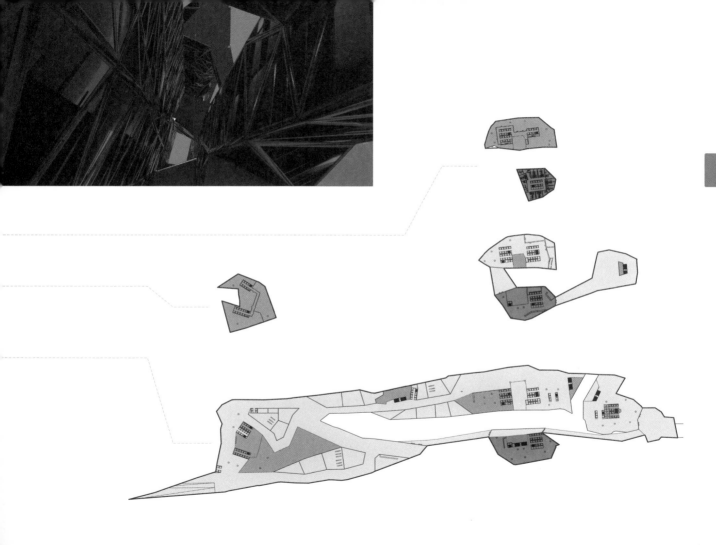

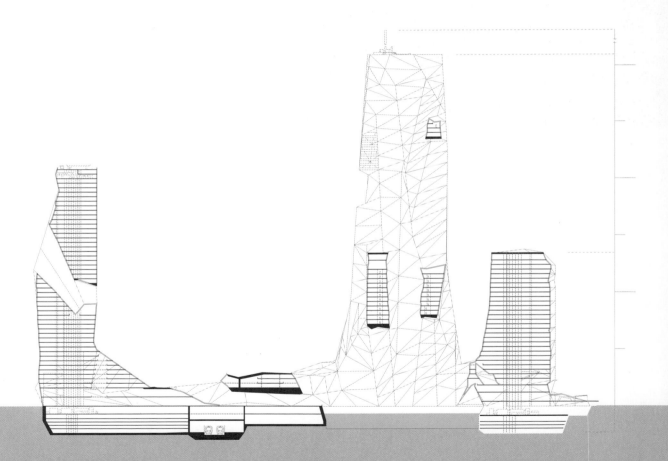

CLUSTER PROJECTS 团队成果

1b

STUDENTS 学生
Krista Glanville «SoA RPI
Chi Zhang «CAUP Tongji

Tectonic Garden
构造花园

In a city subject to constant expansion, the best direction to turn is up. Employing the concept of element as applied to fabric structures, discussed by Irene Emery in The Primary Structures of Fabric, we were able to create a network of multifunctional elements that grew upon itself to become these four towers. Vertical elements attract to each other and apart from one another to create voids that react to site conditions, and interconnect to combine unique towers into one and develop interstitial space between them. Horizontal elements are introduced to direct flow of different traffic scenarios into and through the site at the ground level, becoming housing for public retail space. At higher levels, horizontal elements wind around the towers to create union among them. The array of vertical and horizontal elements and their connections to each other create interiority and exteriority in the realm of space between each interior tower and its individually minded exterior facade.

在一座持续扩张发展的城市中，最好的发展方向便是垂直向上。借鉴Irene Emery在《编织的基本构造》一书中提到的构造原理，我们建立了一个涵盖多功能元素的网络，并依其自身有机发展为四座高塔。垂直元素时而相互吸引时而相互分离，这一过程中建立的空间与基地条件相呼应。这些相互联通的空间在将独立的四座高塔联合为一个整体的同时，还在其间留有大量的间隙空间。水平元素被直接导入地平面以应对不同种类的交通流线，同时演变为公共商业区域。在更高的层面，水平元素缠绕在塔身周围,与高塔形成为一个整体。垂直元素与水平元素的排列分布和它们之间的联系在每一个高塔的内核与外表皮之间形成了室内空间和室外空间。

Tectonic Garden

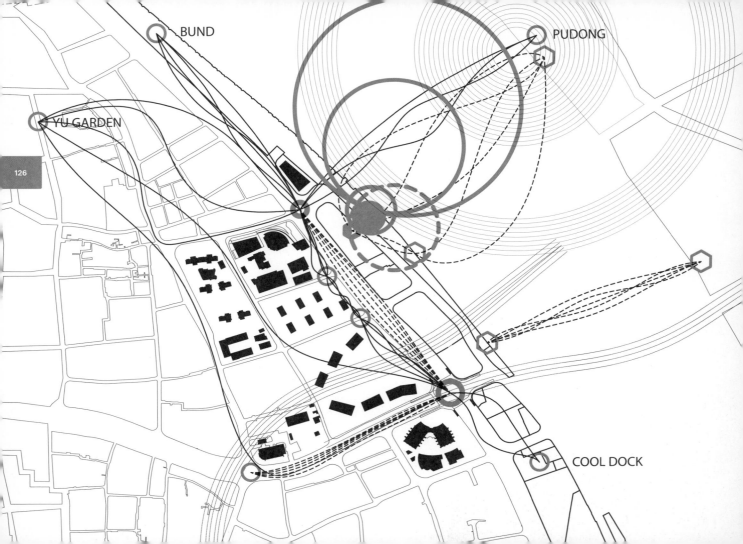

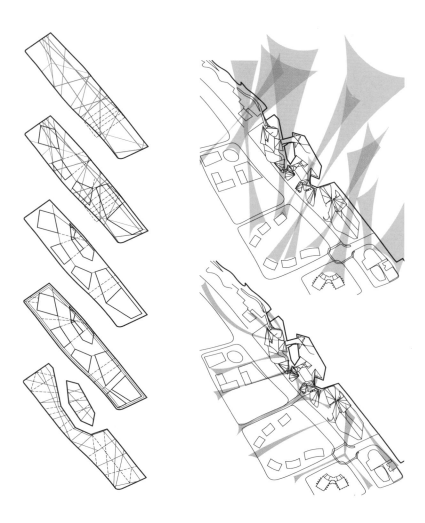

To create the shape of the towers, we used site analysis on motion and views from the surrounding city. This allows us to carve out the river that we brought in to the site to bring the ferry in, and that river becomes an axis for the rest of the site. We continue to carve out the footprints for four buildings that line the new branch of the river, and each building is carved through again to create a cluster of two or three individual towers within each tower that will react with each other in vertical space.

为创造一组高层形状，我们通过周边城市的动态视点来分析基地。这使我们决定切割基地将江水和游船引入基地，引入的河道成为基地其余部分的轴线。我们继续为四栋建筑辟出步道并连接引入的河道，每栋建筑被重新塑性以创造一组由两到三栋塔楼组成的塔楼群，在垂直空间中它们互为作用。

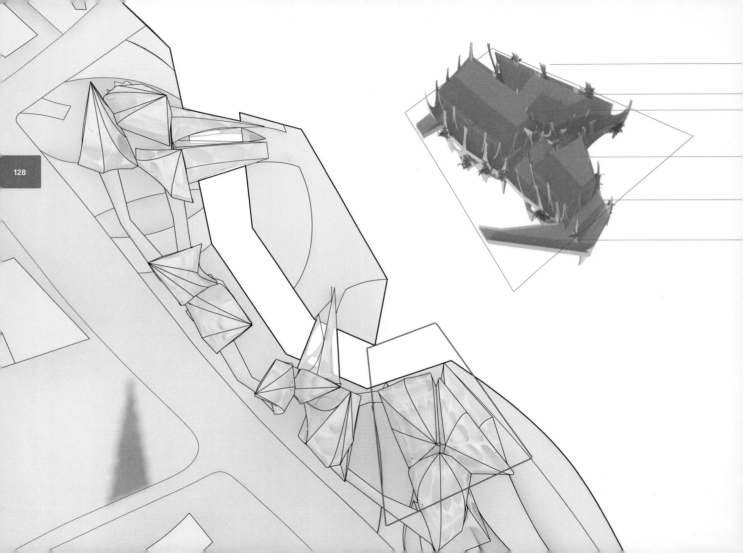

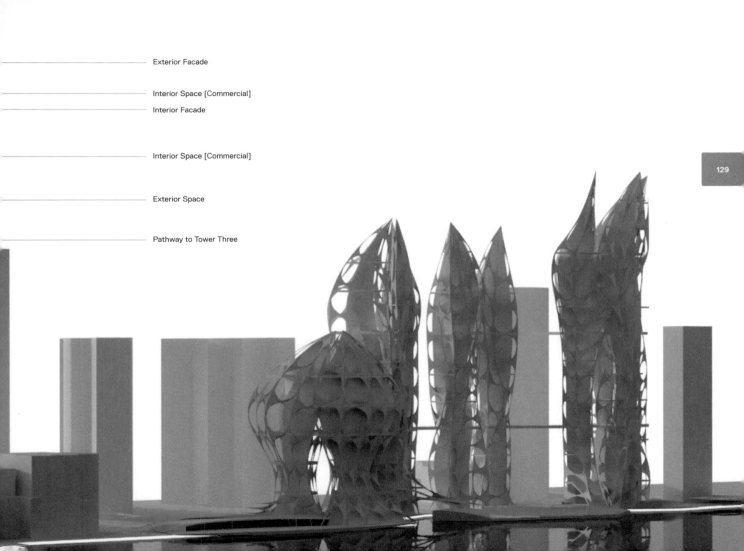

- Exterior Facade
- Interior Space [Commercial]
- Interior Facade
- Interior Space [Commercial]
- Exterior Space
- Pathway to Tower Three

↗ Tower One Unrolled
塔楼一展开

↗ Tower One Perspective
塔楼一透视

↗ Tower Two Perspective
塔楼二透视

↗ Tower Three Perspective
塔楼三透视

↗ Tower Four Perspective
塔楼四透视

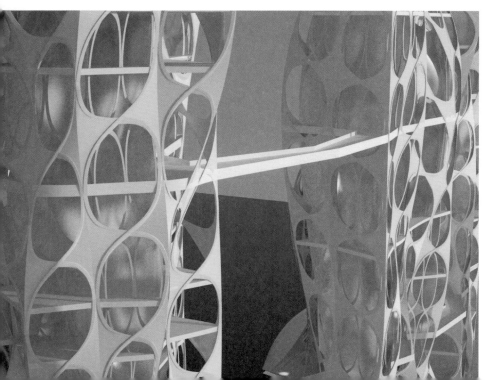

Each tower takes on two boundaries--the interior boundary and exterior boundary. Working with vertical elements that run within the boundaries of these created footprints, we begin to attract each element to each other to allow voids and openings in the forms to react to site conditions. This form beomes the interior boundary of our structure. The exterior boundary was created by the vertical elements that interconnect with one another to combine unique elements into one and create unique interstitial space between the towers that serve as public sky gardens throughout.

每个塔楼有室内和室外两个边界，通过竖向元素的设计延伸了这些有创造性路径的边界，我们开始让每个元素相互间产生吸引并在造型上允许镂空和开放以呼应基地条件。这一形式成了我们结构的内部边界。由垂直元素创造的外部边界相互连接去组合独特的元素并创造独特的在塔楼间的缝隙空间，它们可以成为公共的空中花园。

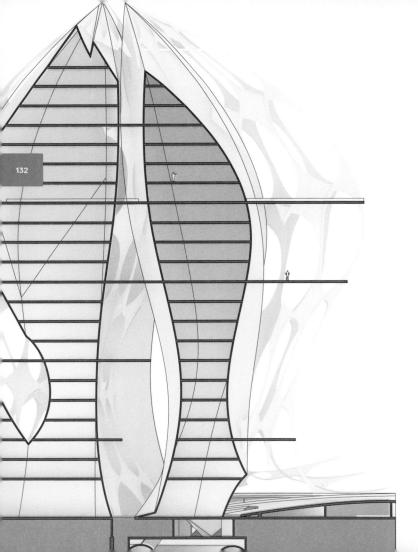

The intent of the exterior facade is to create an open face to maximize the light that would reach the interior floor plates and increase the feeling of openness within the structure, simultaneously incorporating rain capturing methods and housing for vegetation to grow within the wall.

↗ **Facade Detail**
外立面细节

外立面的设计意图是创造一个开放的正面空间，一方面使得室内楼面板的采光率达到最大化；另一方面增加了人们对于整个建筑结构开放性的感受。同时，外墙上也配备了综合雨水收集系统和住房植被系统。

Tower Four includes the entrance to a new metro station. Part of the strategy of the project included creating a new metro stop at the site, which would be crucial to the number of workers and tourists that come to the site for business or leisure.

Like the technique applied to generate forms from the vertical elements that became the four towers, horizontal elements take on a similar profile to create pathways into the site and create horizontally spanning retail space.

四号塔同时也包含了通向新地铁站的入口。在基地上包含一个新建地铁站是项目策略中的一个部分。这对来到该地区休闲和商务的工作者、旅游者都很重要。就像从垂直元素运用技术生成外形而形成的四座塔楼，水平元素采取了类似的办法去创建路径进入该基地，并创建水平跨越的零售空间。

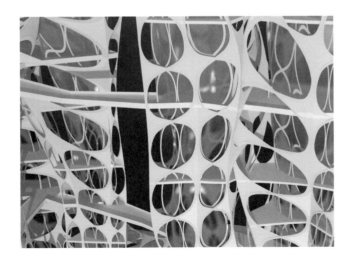

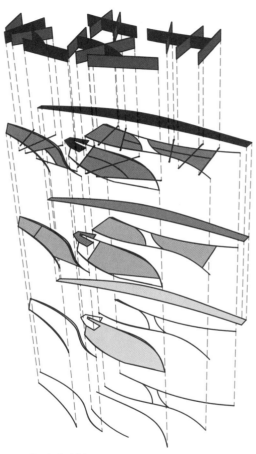

↗ Exploded Diagram

2ª

CLUSTER PROJECTS 团队成果

STUDENTS 学生
Desireé Edge «SoA RPI»
Houzhe Xu «CAUP Tongji»

Twisted Wrap Through Branch Connection
分支连接上扭曲的包裹

South of the Pudong area, the dwindling territory of green space is becoming an issue worth discovering. By emphasizing the bringing together of two different styles within the existing city, the idea was to make a direct link to both mediums. In order to do so, making direct and visual continuous connections within and outside of the site to not only provoke a conversation of making relationships between the existing context, but to also evoke a new experiential dialogue between different levelled programs was imperative. This importance was a direct observation of the limited interaction the existing city has with the Huangpu River and the dispersion of green space throughout. With the connections, a system of wrapping and twisting structural members to frame the key levels of the design, allowed for a multi-view interaction. By having the elevated hotel above the entertainment and commercial areas, an increased interaction with the bottom levels was provoked to bring more movement and functionality to the upper levels.

在浦东南区，绿地环境的不断减少是一个日益严重、值得关注的城市问题。通过将城市中现有的两种模式结合在一起，我们希望在它们之间建立一种直接的联系。为此，必须采取直接的、视觉上通畅的手段连接基地内部与外部的空间节点。这样不仅可以激发项目与既存城市文脉之间的对话，更可以建立不同功能层之间全新的互动体验。这是对既存的城市空间缺乏与黄浦江、零散绿地的互动这一现状的直接观察而得出的。通过建立连接，一套覆盖、扭曲的结构系统上升为设计的关键点。通过将宾馆布置于娱乐、商业功能区之上，建筑的高层与低层的互动将更加富有活力，有助于低层使用活动和功能的向上渗透。

Twisted Wrap Through Branch Construction

黄浦江与周边道路主要连接点的基地图示
Site diagrams of major connection points off the Huangpu River for circulation avenues.

By identifying key connection points outside of the context, the design makes direct connections to the outside attractions as a way to draw movement into the site. Through interlinked circulation lines, the design creates a sense of direction at multi-levels that interwove into itself to make direct connections at different levels.

The points identified outside of the site, not only direct the start of the wrap, but also the points of location for the design to connect on the ground with the existing road system in order to provoke a dialogue with the ground plane and the plane above the road.

通过确定与外部环境的关键联系点，设计建立与外部有吸引力区域进入基地的直接联系，通过相互循环流线，设计创造了多层次的方向感：把四通八达的路径交织起来，在不同层面上形成了明晰的路径。

在基地外确定的连接点，不仅承担着表面上的起点功能，同时也是现有的道路系统连接基地地面的设计联系点，从而与地面和空中的平面建立起对话。

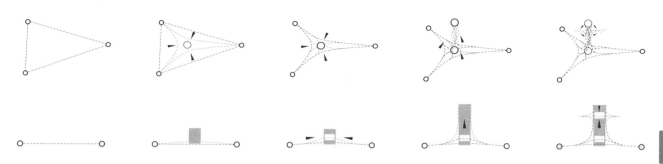

Major Attention Points » A triangulated systems that pulls up vertically where the elevated park intersectes the towers for vertical circulation and massing configuration.

主要关注点 » 被三角形系统垂直撑起的高架公园，与塔楼相交织，形成了垂直流线和集中分布。

高架公园和地平面的联系
Connections between the elevated park and the ground plane.

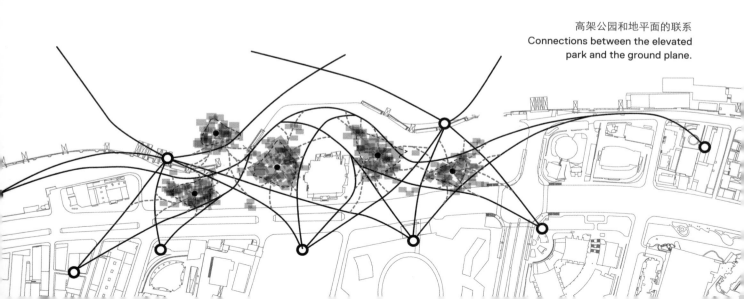

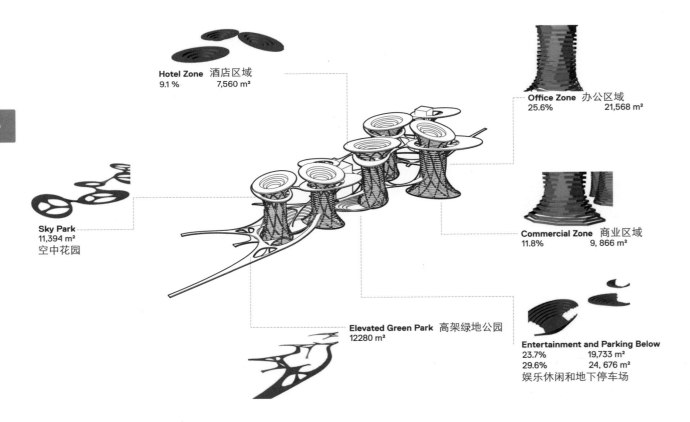

Programmatic Diagram » Locations of various programmatic functions and activities throughout the project proposal.

功能图表 » 项目设计的各种功能和活动

↗ **Angles of view from various positions within towers.**
塔内不同位置所得到的视野角度

Twisted glass structure relationship to woven stucture.
扭型玻璃结构和交织结构的关系

外部环形广场与高塔基座的关系
Condition created between outdoor amphitheater and tower base.

建筑外皮交织结构的细节
Detail of enclosure to intertwined structure.

从空中酒店空间通往高塔的细节
Detail of elevated hotel space to tower.

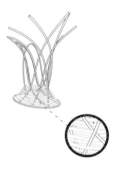

Tower 1

Tower 2

Tower 3

Tower 4

Tower 5

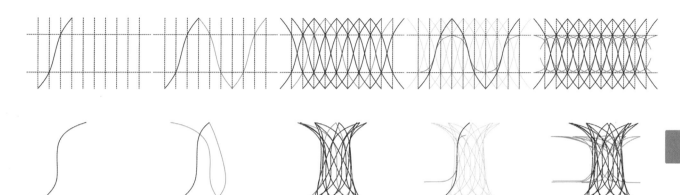

Weaving Pattern » Illustrates the use of a single element connected by twisting across the plane and wrapping into a circular massing.
编制样式 » 图解了用一个单一的设计元素通过平面扭曲和包裹形成一个圆形的体块。

Based off of the urban strategy of connecting to attractions within the context, the towers were rotated to allow for a woven motion in and out of towers for enhanced views from ground and elevated levels. This wrapping motion was carried into the twisted structure and enclosure as well which allowed for connections of the retail amphitheater locations and the sky park connections to the towers.

基于城市策略中的景观与环境的联系,所有塔楼运用内外编织运动被旋转,从而增强地面与空中的景观视野。这个包裹型的运动同时也运用到了扭型结构和建筑外皮结构中,使环形商业广场和空中花园都建立了与塔楼的联系。

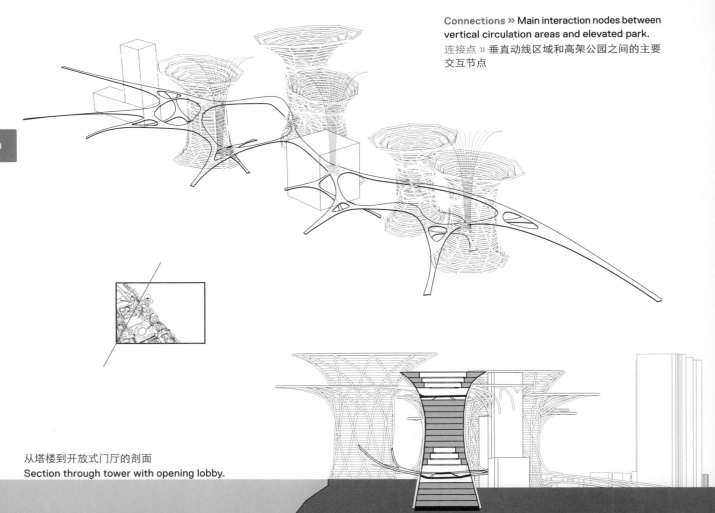

Connections » Main interaction nodes between vertical circulation areas and elevated park.
连接点 » 垂直动线区域和高架公园之间的主要交互节点

从塔楼到开放式门厅的剖面
Section through tower with opening lobby.

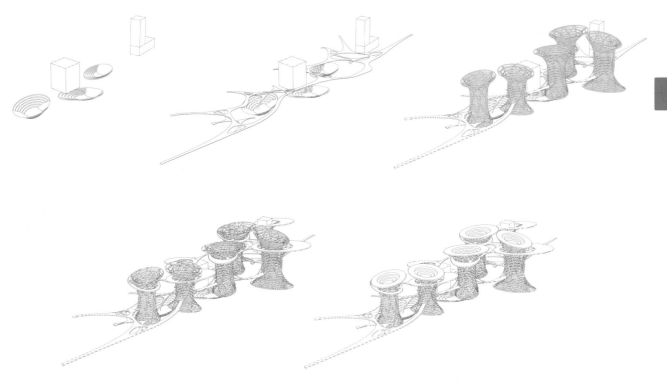

体块 » 主要项目空间的主要体块
Massing » Main massing of major program spaces.

2b

CLUSTER PROJECTS 团队成果

STUDENTS 学生
Michael Stradley «SoA RPI»
Yifeng Wu «CAUP Tongji»

Form and Performance » Deep Surface
形式和性能»深层表面

While the logics of the individual loop and knot are simple, their structure allows them to tessellate into highly complex and varied assemblies. Within the context of the parametric project in architecture, these primary structures yield novel approaches to form and space. The operation of weaving, decoded and re-coded through the physical and drawn diagram, begins to offer inventive solutions to the concerns facing architecture and urbanism within contemporary Shanghai. Entering into dialogue with the complexity of Shanghai today, the proposal seeks to reconcile the speed and scale of the new Chinese city with the moments of community and intimacy that so characterized the traditional Chinese lilongs. The project explodes the urban fabric into several horizontal strata and vertical connective nodes. As a critique of present-day Chinese urbanism, the proposal attempts to re-read aspects of historical Shanghai and re-insert moments of cultural memory into the city via architectural intervention.

尽管在编织中单独的环和结自身逻辑相对简单，但其结构特点使得它们能够相互交织，形成高度复杂而多样的集合体。在参数化设计的帮助下，这些基本结构便可以生长衍化出新颖的造型和空间。编织的过程为应对当代上海建筑与城市规划中的诸多问题提供了一种创造性的解决方案。在与如今上海极度复杂的城市现状的对话中，本方案试图在典型的中国城市的高速度生活和巨型尺度与上海传统里弄的社区空间和邻里生活两者之间中找到一种平衡。本方案将已有的城市构架拆分为一系列水平向的层，并建立相应的垂直联系节点。作为对当代中国城市规划方式的一种反思，本方案尝试对上海历史进行再解读，并通过建筑的手法将文化记忆的瞬间重新植入城市,带来新的活力。

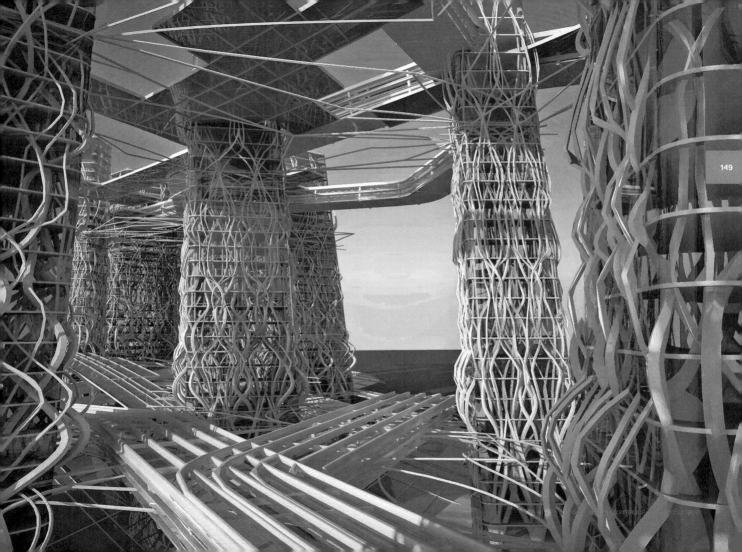

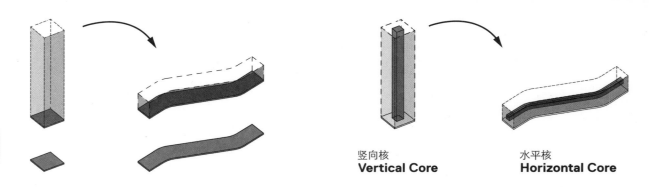

竖向核
Vertical Core

水平核
Horizontal Core

Situated south of the historical Bund in an area of high traffic, the project responds to existing infrastructures by establishing two elevated ground planes. Drawing from the classical Greek temple, the plinth exists to negotiate between architecture and ground, in terms of both the structural and programmatic organizations.

因为坐落于交通繁忙、历史悠久的南外滩，本方案通过建立了两个升高的地平面来回应现有的城市基础设施。设计借鉴古希腊神庙柱基与地面和建筑的关系转换成结构与方案组织相互关系。

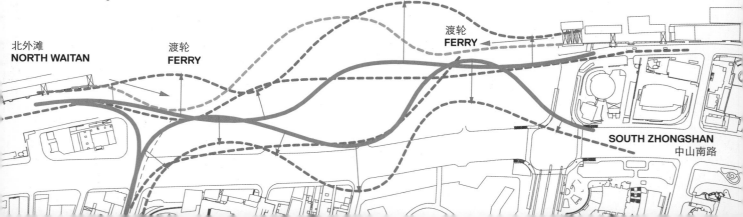

This logic is mirrored, establishing two inhabitable landscapes, one of ground and one of sky. From the plinths emerge a number of vertical elements, a massive-scale inhabitable column screen that links the upper and lower public parks. These publically negotiable plinths rise above the street level as floating parks, accessible only by pedestrians and cyclists. Automobile and public transportation remain at the existing street level. The existing ferry terminals are moved slightly inward, allowing access to ground level traffic (high speed) as well as the new elevated level traffic (low speed).

本方案的思路是映射，建立两个可居住的景观地，一个在地面、一个在空中。从柱基呈现的诸多垂直元素，形成了一个大规模的可居住的柱形建筑群，同时这些柱群也连接着高、低两个公共公园。这些公共的柱基从街道升起像漂浮的公园，只有通过人行道和自行车道才可以进入。汽车和公共交通仍然使用现有的道路系统。现有的渡轮站点被少许移动至内部，使地面交通（高速）与新的空中交通（低速）都可以进入。

方案组织的剖面图示
Sectional Diagram of Program Organization

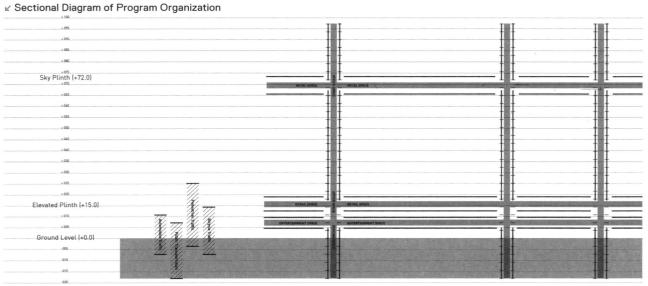

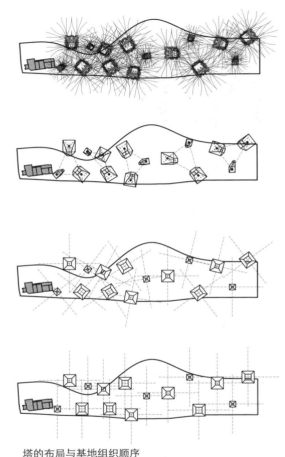

The overall configuration of the towers is generated by a series of operations on the site aimed at generating unique spaces and conditions between the towers. Diverging from a regular layout, the towers defy the regularity of the grid and undergo orthogonal slippages. From there, the towers recognize smaller groupings with other towers, rotating to face the space between them. The towers then lean inward over the courtyard space formed by each group. In these new positions, the towers witness the deployment of a woven façade system, linking them further.

所有高塔的总体分布是基于整个基地的一系列设计而生成的，目的是创造塔与塔之间特别的空间和联系。虽然源于常规布局，但高塔的设计不受常规网格和正交滑移的定式思维局限。因此，高塔之间形成了更小的组群，旋转面对它们之间的空间。然后，高塔会倚靠在各自组群形成的庭院区域中。在这些新位置上，所有高塔都面对着各自的编织外立面系统，使它们的联系进一步加强。

塔的布局与基地组织顺序
↗ **Tower Placement and Site Organization Sequence**

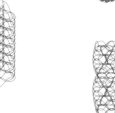
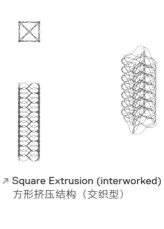

↗ Square Extrusion (interworked)
方形挤压结构（交织型）

↗ Torsional Circle (interworked)
扭转圆（交织型）

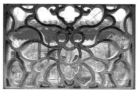

↗ Square Extrusion
方形挤压结构

↗ Torsional Circle
扭转圆

↗ Screen Patterns
幕墙样式

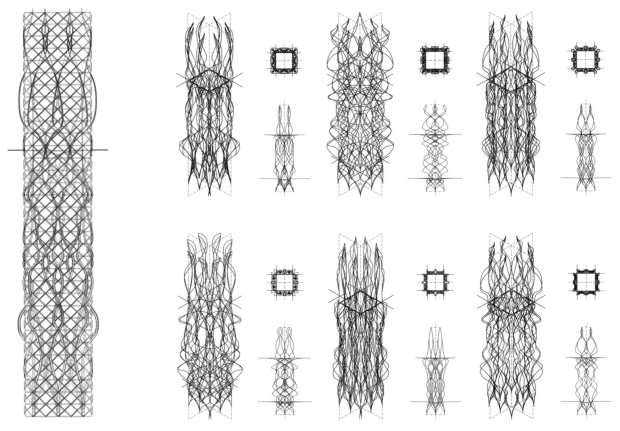

原型立面
↗ Prototypical Elevation

表皮建构模式群组
↗ Family of Tectonic Veil Formations

It is no coincidence that the weaver's lexicon shares much with that of the parametric architect. Operations performed by the digits (fingers) for hundreds of years now enter the space of the computer as digits (numerical values), performing the same processes with a deep structure made of zeroes and ones rather than physical operations. Essentially performing as structural skins thousands of years before the architectural interest in the term, the practice of weaving synthesizes form and structure through the manipulation and multiplication of a tectonic logic.

Conceived of as generative machines within the space of the computer, diagrams evolve into positive feedback loops of content and form. A diagrammatic examination of performance, in this case, the performance of fabrics, transcends its inputs, transforming an analysis of specific tectonic phenomena into architectonic structures laced with spatial and cultural intelligence.

参数化建筑师从编织者的词汇里分享了很多,这绝非偶然。几百年来设计师们一直由数字（手指）完成的工作,如今进入了计算机空间,数码（数值）代替了人力,它使用0、1深层结构程序完成了以前只有人力才能胜任的类似工作。在受到建筑界重视以前,编织的运用通过建构逻辑的操纵和增加,综合了形式和结构,作为结构表皮的本体发展已经有几千年。

设想有着计算机空间的生成性机器,图表进化到了内容和形式的积极反馈循环。在这种情况下,一个程序式的性能检查,织物的表现力超越了它的原材料,使一个具体的建构现象分析转变为了用空间和文化理解力作为装饰的建筑结构。

↗ Aerial Siteplan

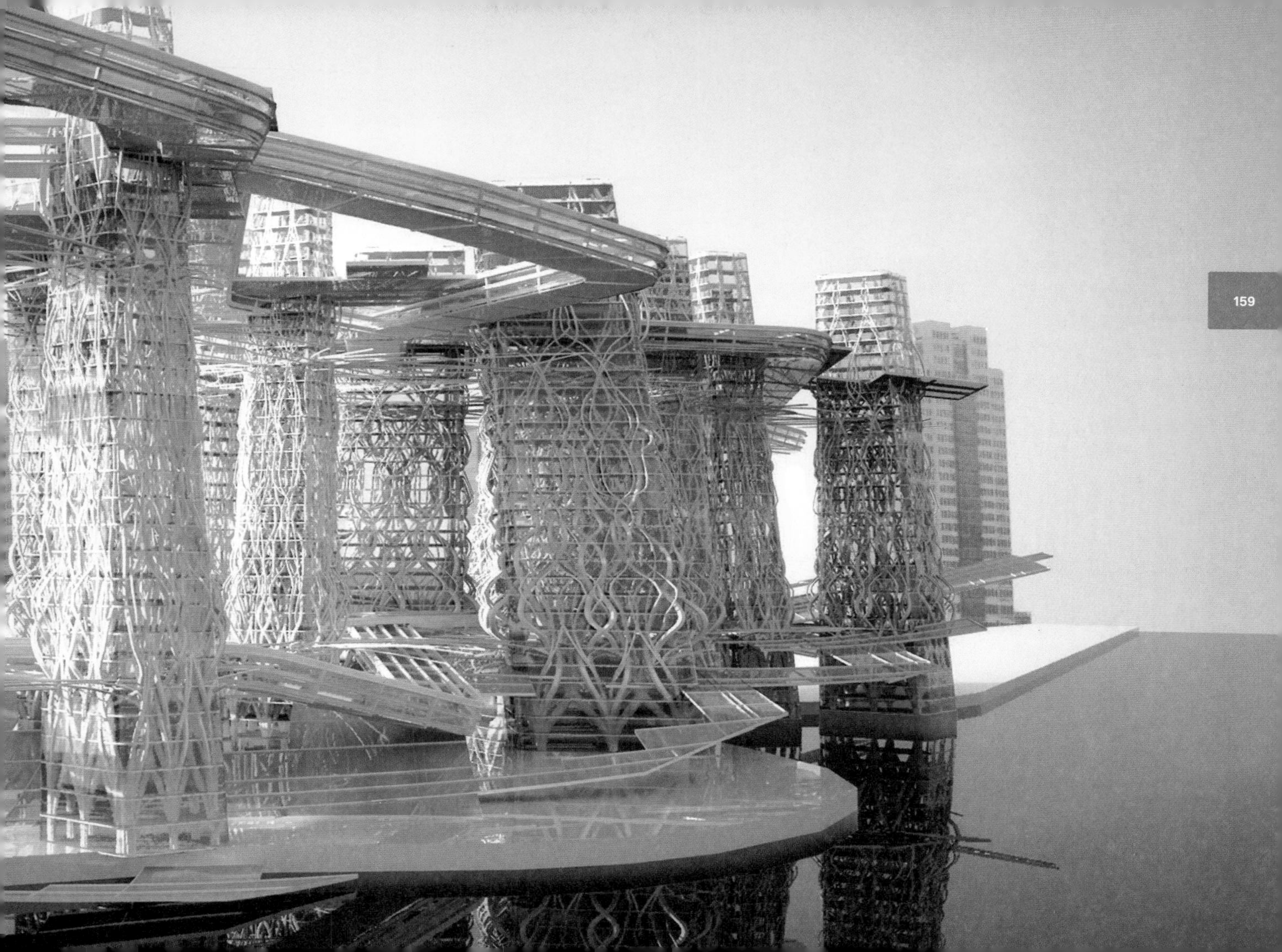

3ª

CLUSTER PROJECTS 团队成果

STUDENTS 学生
Seth Hepler «SoA RPI
Jun Su «CAUP Tongji

Good People Happy Land Building
幸福市民，景观建筑

Critique of modern urbanism often focuses on architectures lack of relation to the urban fabric as it pertains to the human scale. Working in a site that easily could fall back to architecture for the vehicle, a conscious move is needed to separate the notion of building and site in an attempt to refocus the architectural intentions on a space tailored to the human scale. Surrounded by contemporary urbanism, the project hopes to serve as a symbol, an icon of intention. Urbanism must be stolen from the grips of the automobile and given back the people. Our role as architects goes beyond simply creation of novel spaces and buildings that respond to human needs. We are responsible for the impact that modern urbanism is having our the earth and our lives. Therefore, a new urbanism that is focused on taking the lessons from the past and reinterpreting successful elements from them into a modern typology becomes the role of the architect in this day and age.

当代的城市化因为建筑与人的尺度失衡这一问题而常常被指责缺乏与城市肌理的联系。在一个很容易就变成为汽车而设计的场地里，我们需要有意识地区分建筑与场地的概念，试图重新调整到适合人体尺度的建筑空间。本方案希望在当代都市环境的包围之中能成为一个符号，一个强烈意愿的图标。在过去的规划中被汽车所偷走的城市空间必须还给市民。我们作为建筑师不仅仅要回应人们的需求而设计新的空间和建筑。我们手中掌握的是现代城市对地球和我们生活的巨大影响。因此，今天的建筑师必须从历史中学到经验教训，从中获取有价值的元素，并将其转换成为一个现代化的城市类型，为人们所用。

↗ Solid-Void Diagrams
图底关系

悬空于地面、就像脱离了地球引力一般，本方案设计了一个空中柱形建筑，既从实体的和意识上与当前上海的都市化隔离开，也在都市环境下从人的尺度构造了一个真正的现代化设施。

基地的边界被挤压成简单的板，并在空中被很多腿贯穿。基地上已存在的板作为公共的柱基是外滩的直接延伸，也是公交车、汽车、渡轮、地铁新的交通枢纽。

Divorced from the ground, and seemingly from gravity, the projects operates as an elevated plinth that both physically and ideologically provides a separation from current Shanghai urbanism and creates a truly modern manifestation of human scale in the urban condition.

The site boundaries are extruded into a simple slab which is then pentrated by the feet of our city in the sky. The slabs existence on the site serves as a public plynth that is a direct extension of the Bund and as the new trasportation hub for bus, car, ferry, and metro.

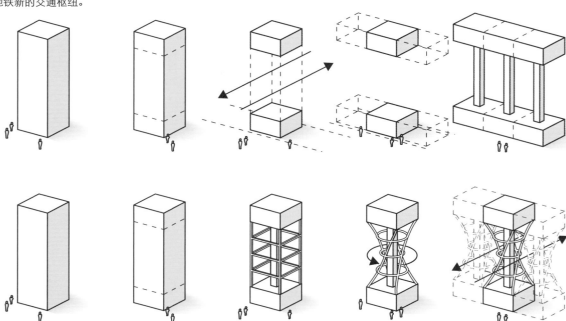

↗ Formal Generation 形式生成

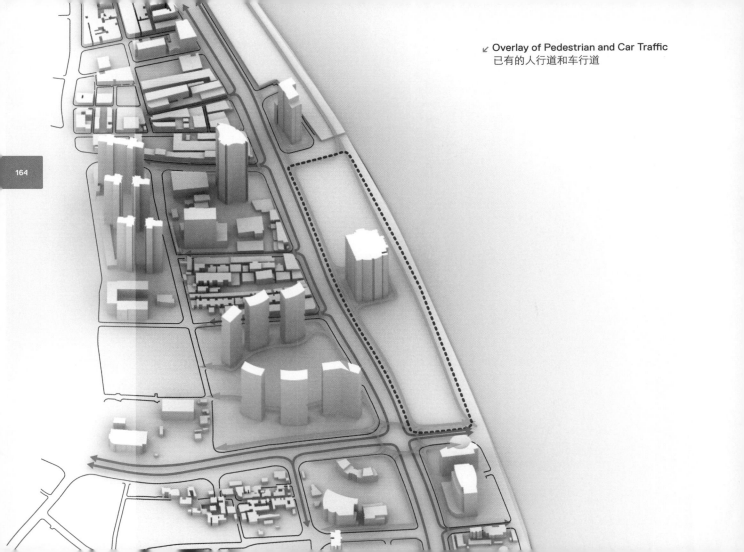

↙ Overlay of Pedestrian and Car Traffic
已有的人行道和车行道

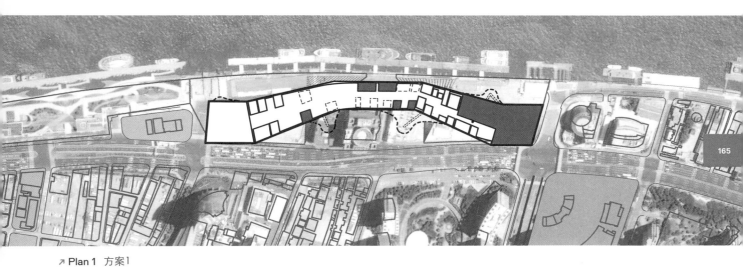
↗ Plan 1 方案1

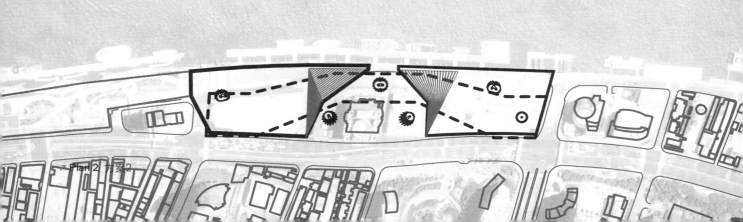
↗ Plan 2 方案2

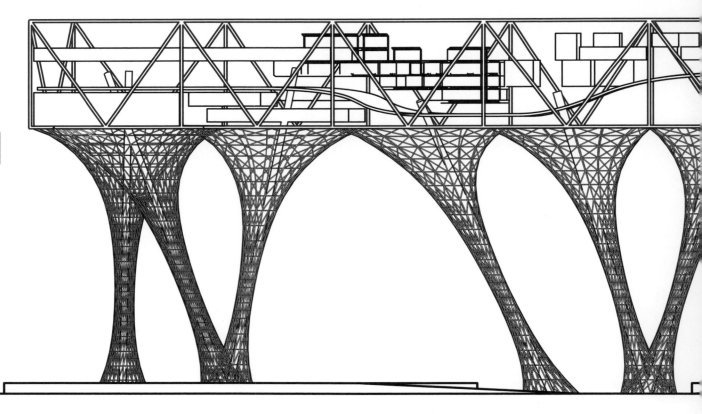

The parti of the building speaks mostly to the elevation or section. The project vertical circulation between the public plynth and transportation infrastructure on the ground and the public plynth and program spaces in the upper bar is a series of angled elevator and stair cores.

建筑主体更多强调竖向性和剖面性。方案的垂直动线在公共柱基和地面交通基础设施之间，公共柱基和上面长条功能空间之间是一系列成角度的电梯和楼梯。

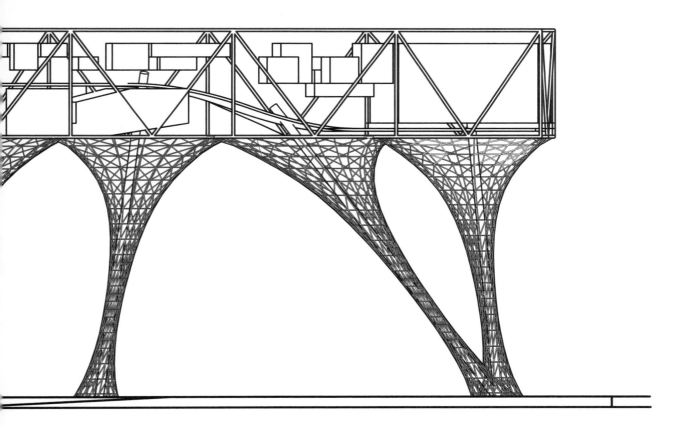

CLUSTER PROJECTS 团队成果

3^b

STUDENTS 学生
Anita Rodgers ‹SoA RPI›
Jialin Yuan ‹CAUP Tongji›

A Fabric of Volumes
编织体量

Traditional basket weaving is made of a series of overlapping strands to develop a flat fabric. This project explores the technique of basket weaving, but with a curved element. The curved element creates a new logic in the "fabric that is created. The fabric is not flat, but consists of a series of volumes whose scale is determined by the radii of the curved elements that create it. The project aims to re-invigorate the middle of the skyscraper by removing it and giving back to the public as a uniquely vertical public space. The building with the middle removed is a building freed from the ground. The physical division of the top of the building, allows it to explore the site at human scale, while the bottom part still remains embedded in the city. The weaving system defines spaces on the ground that serve the top as lobbies. A small taste of what is above on the ground to encourage people to explore the site vertically, and discover what is above.

传统的编织竹篮是由一系列的线交叠而成的平面织物。本方案同样从竹篮编织出发，但是加入了曲线的元素。曲线单体在织物中建立了一种新的逻辑。织物不是完全平整的，而是由一系列空间体量所组成，其尺度由曲线单体的半径决定。本方案通过去除高层建筑中间的部分并建立垂直向的公共空间来改善建筑的中部，解放而独立于地面。建筑上部体块的空间分割划分了场地中人的尺度，而下部则归于城市尺度。编织体系在地面高度所定义的空间作为首层大厅服务于上层部分。在这里公众将得以感受部分上部空间而被引导向上。

A Fabric of Volumes

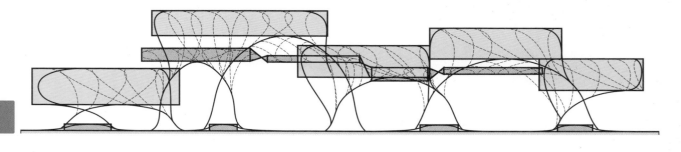

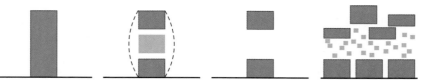

↗ Programmatic Relationships

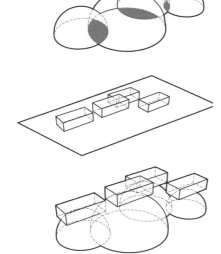

Conceptually, the middle of the building has been removed, which requires it to physically be mostly empty. However, on the ground the project requires a small amount of program space to serve elevator cores and lobbies for the hotel and offices, which necessitates two types of spaces: the Fullspace (where lobbies and circulation are located), and Voidspace (which are the empty open spaces outside of the Fullspace). Consequently, a system was created to locate the Fullspaces along the site so that they can be enclosed by the skin of the building, and, at the same time, leave open most of the space at the bund-level landscape, as Voidspace for the public to use. Because the waeving logic forms slightly intersecting volumes, it naturally cretes two sets of spaces. The spaces enclosed by one volume become Voidspace while those enclosed by two volumes become Fullspace.

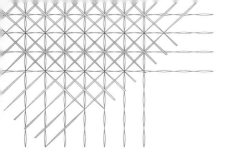

从概念上来说，建筑的中心部分被移走了，使之在实体上空出大部分空间。然而在地面上，本方案还需要一些功能空间用以放置酒店和办公区的电梯核心筒和大堂。这也形成了两种意义上的空间：实体空间（放置大堂和交通）和中空空间（在实体空间外的空的开放空间）。因此，这个空间系统既可通过被建筑外壳的包围来用于定位该区域内的实体空间，亦可同时在外滩标高地景留出大部分的中空空间供公众使用。因为编织逻辑微妙的形成了交叉的实体，很自然地创造了上述两种空间。被一种实体包围的空间是中空空间，而被两种实体包围的成为了实体空间。

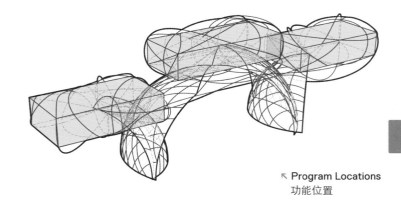

↖ **Program Locations**
功能位置

↗ **Dome Volumes Before Splitting**
裂开前的穹顶容积

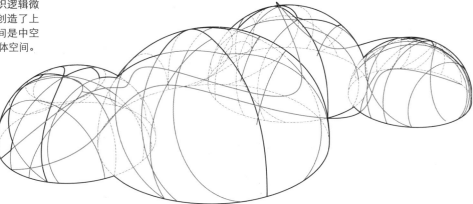

↗ **Sectional Programmatic Relationships**
剖面的功能联系

在地面上，本方案还需要一些功能空间用以放置酒店和办公区的电梯核心筒和大堂。然而，位于中央的中空空间对于本方案的核心理念至关重要。由编织系统推进的功能和实体是由每个区域的元素数量以及这些元素在剖面上的尺度如何互相转化来决定的。所以设计的这些实体都是一个大型网络中的一部分，但有基于实体的样式和概念深入的不同编织密度。最终，这些实体形成的网络清晰可见，一些植根于地面，另一些则连接起其他的实体。

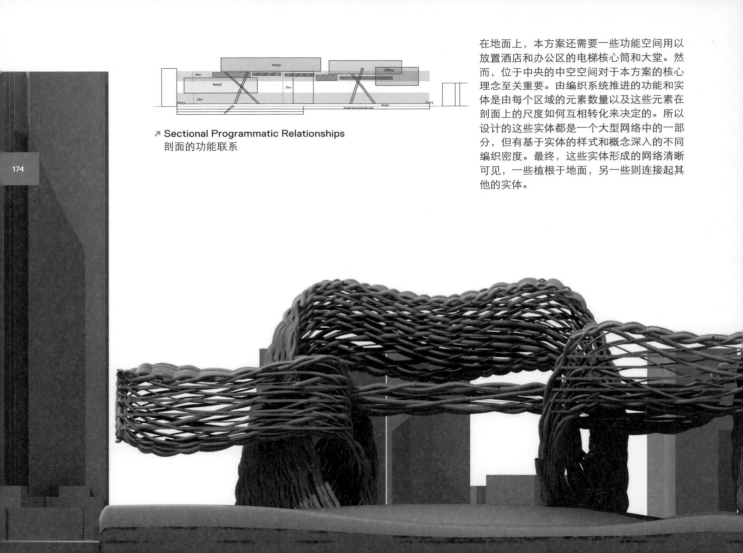

On the ground the project requires a small amount of program space to serve elevator cores and lobbies for the hotel and offices. However, the voidspace in the middle is very important for the clarity of this project's concept. The progession from program and massing through weaving system was determined by the number of elements in each area and how they move as they transition from one scale to another in section. So that the volumes that are created are all a part of the larger network, but have different densities of weaving based on the pattern of volumes that permeates the project. The result is the network of volumes that can be seen below, some which are tied to the ground, and some which, as the core, are connected only to other volumes.

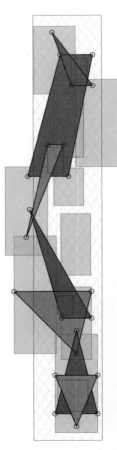

↗ **Viewports**

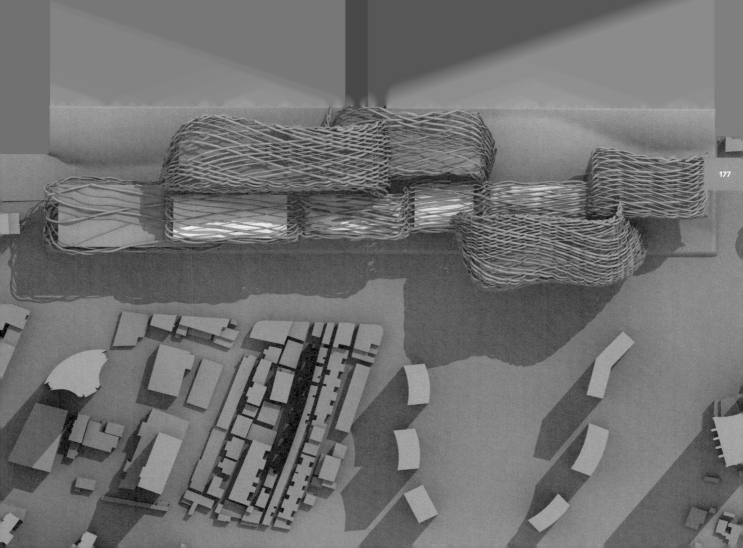

178

4ª

CLUSTER PROJECTS 团队成果

STUDENTS 学生
Jason Wang «SoA RPI»
Wenqian Jiang «CAUP Tongji»

Shanghai Link
上海链接

Shanghai is a city that has been sculpted over the past century as a result of occupying countries and international trends. These infusions leave behind a trace that can be found throughout the urban fabric; from it emerges a distinct separation between cultures, ideals and classes. Shanghai Link responds to this separation by instigating an interaction between the fundamental factors that built this split people. Commuters, Residents, Tourists, and Businessmen are funnelled into our site by the use of the public transportation systems, and then are encouraged to stay as they wander through a multitude of public areas. The ground plan is manipulated around main circulation arteries and the shape and height is determined by visual interaction within the space. An investigation of a visual linkage between Puxi and Pudong generates viewing points that are spread throughout the site that influence the form of the high rises. The atriums have also been shifted towards the exteriors in order to increase visual interaction.

在过去的一个世纪中，上海因各国分区租借和国际贸易的影响而形成了独特的城市文脉，文化、意识和阶级均被明确地分化了。上海链接试图回归造成这种分化的最根本因素——人，通过建立他们之间的互动来修复这个问题。通勤者、附近居民、游客、白领由公共交通系统引导入基地，并且在一系列在地面上起伏的公共空间中逗留。底层平面沿主要空间流线布置，其形状和高度由空间内视觉焦点的交互来确定。浦西浦东间视觉通道建立的一系列场地视点影响着高层建筑的整体造型。边庭的布置也创造了更多的视觉互动。

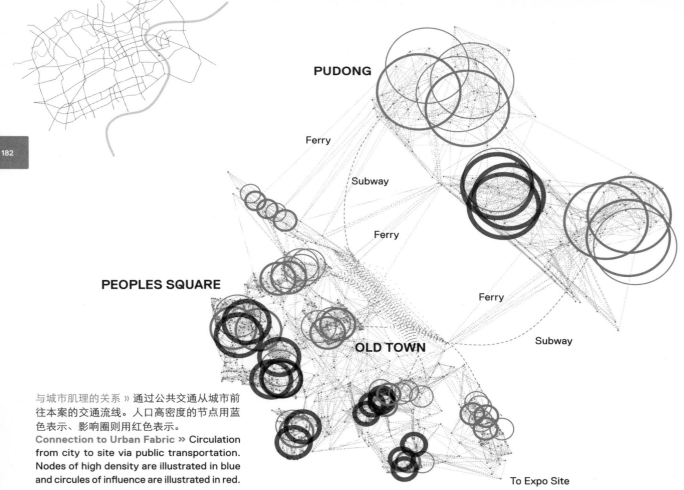

与城市肌理的关系 » 通过公共交通从城市前往本案的交通流线。人口高密度的节点用蓝色表示、影响圈则用红色表示。
Connection to Urban Fabric » Circulation from city to site via public transportation. Nodes of high density are illustrated in blue and circles of influence are illustrated in red.

In order to begin to understand our site we began with a study of relationships in Shanghai. By mapping transportation routes, points of interest, commercial areas, residential areas, public space, and business areas the relationships between different elements begin to become more apparent and densities begin to appear that are a clear indication of a rift in the urban fabric. Using this approach we carefully designed our site to suit the connections through the site.

Based off our initial studies of transportation routes and population densities the more effective solution to draw people into our site without program would be to alter the near-by forms of public transportation so that they would be situated in our site or that our site would have an access point.

为了更好地理解我们的基地，我们首先从上海的关系研究来入手。通过绘制交通路线图、兴趣点、商业区、居住区、公共区域和商务区域，使这些不同元素之间的关系变得更加明显，密度也呈现出一幅在城市肌理中的一叶孤舟的景象。通过这个方法我们谨慎地设计我们的方案，以便使本案和当地各地区的联系相适应。

基于我们先前对于公共交通和人口密度的初步研究，如果没有计划，吸引人们来到本案的更有成效的解决办法就是改变周边的公共交通的形式，使它们适应本基地或者使本基地拥有一个交通进入点。

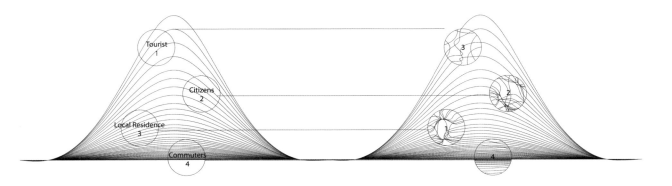

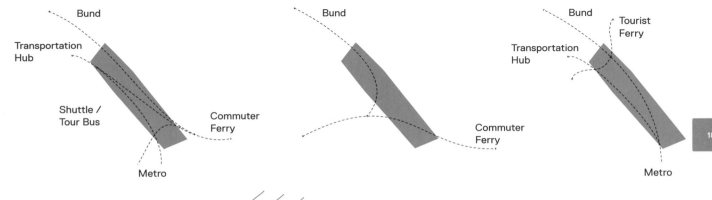

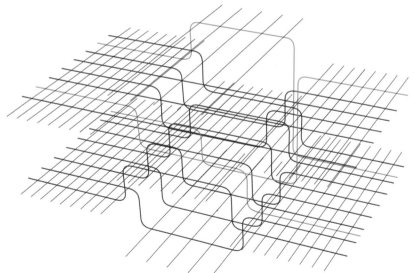

As a result of altering access points to our site new diagrams were created in order to re-evaluate the new circulation paths throughout our site and to determine which conditions would provide the most interaction amoung different people.

作为更改通往本基地的交通进入点的结果，我们制作了新的图表来重新评估贯通本基地的全新交通流线，同时也决定了在不同人群之间，哪一种情况能提供最高效的互动。

The forms are generated through a series of steps that question the regular arrangement of buildings. Atriums, high rises, and the ground condition are all re-evaluated in order to maximize an interaction between distinct groups of people.

An atrium is a central hall or open courtyard in a building. The idea is to increase visual interactions within the building. However, our goal is to expand the area in which people are able to interact whether it be physically or visually. For example, by shifting the core to the exterior of any given floor the views instantly begin to expand out. Instead of only viewing people within a single building, an individual may begin to view into surrounding buildings therefore creating visual connections where there were none before and thereby creating an exchange.

A system of raising and lower the ground plane increases the number of both physical and visual interactions occupants will have with people throughout the site as well as having the opportunity to view through our site.

本案的设计通过一系列步骤逐渐生成，对传统的建筑布局提出了新的见解。中庭、高楼和地面环境都经过了重新评估，来使不同群体的人们获得最大限度的互动。

中庭指建筑中的中央大厅或者开放式庭院。我们的灵感在于增加人们与建筑之间的视觉互动。然而，我们的目标是拓展人们的互动范围，无论是物理上还是视觉上。举个例子，把任何一个楼层的核心转移到外部去，就可以使景观视野持续地增加。人们不再是在单一建筑进行眺望，取而代之的是可以全面地环视周围所有建筑，从而建立了视觉上的联系和转换。

一个使地平面上升或下降的系统增加了居住者在本案物理上和视觉上的互动，同时也有机会从本案极目远眺。

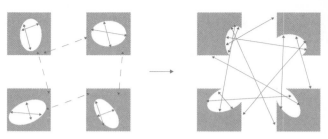

Atrium Interaction » Visual interaction between individuals in a standard atrium versus the proposed project's atrium spaces.
中庭互动》基本中庭的个体与预定计划中的中庭空间之间的视觉互动

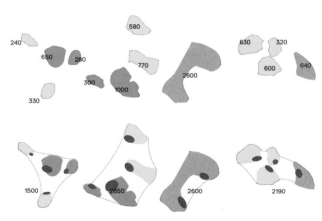

Atrium Metrics » Connection and expansion after atrium program is moved to the edges
中庭韵律》中庭方案被移动到边缘后的联系和扩展

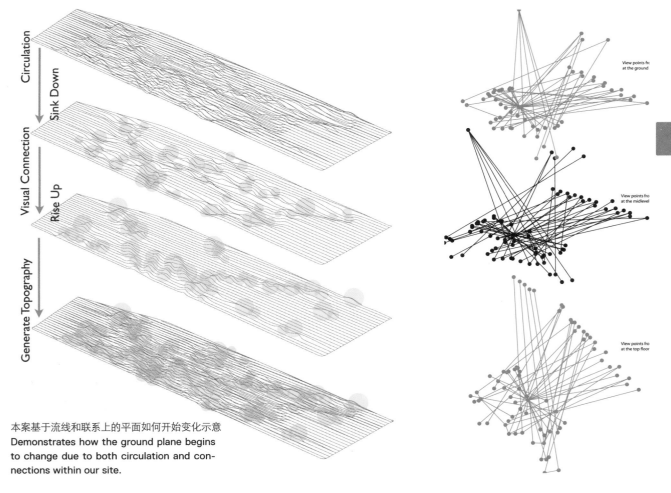

本案基于流线和联系上的平面如何开始变化示意
Demonstrates how the ground plane begins to change due to both circulation and connections within our site.

The ground plan steps up and becomes floors to provide opportunity for retail and other program geared towards residents and tourists. Further up studios and a museum become the main program and attempt to draw people up vertically throughout our site. In order to facilitate this vertical movement vertical extensions of public space and gardens allow pedestrians and tourists to migrate upward without interfering with more private programs. The water is slightly pulled into our site in order to house a small pier.

地面计划逐级而上，成为一个供商业和居住、观光等其他功能的平面。更进一步，工作室和博物馆成为了主要的功能，也使人们垂直向上穿过我们的建筑群。为了促进这种垂直移动和垂直扩展，公共空间和花园使行人和游人向上移动，以免打扰到更多的私人空间。水被轻微地引入本基地以形成一个码头。

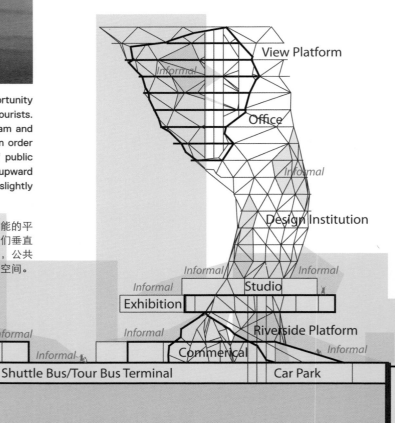

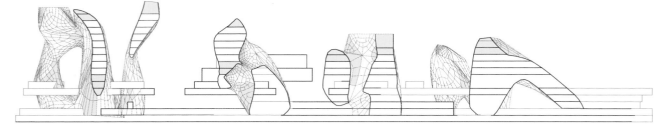

Programmatic Distribution » Plan cuts through the tower illustrating the distribution of program throughout.

功能安排 » 表达功能安排的塔楼平面

191

CLUSTER PROJECTS 团队成果

4b

STUDENTS 学生
Travis Lydon «SoA RPI
Bin Long «CAUP Tongji

Renegotiating Urban Density
重新调和城市密度

Utilizing an systematic approach to design, the project acknowledges observed densities within the urban fabric as a base for analysis. Utilizing the surrounding context as an initial point of departure, the system is based on inter-relations between physical instances within the built environment, re-establishing certain focal zones or 'nodes' within the fabric of Shanghai. The system establishes points of departure within the system and, based on adjacent densities and degrees of separation between points of influence, utilizes a very organic growth pattern which define a general volumetric space. Programmatically, the intervention and insertion of much more publically-driven program begins to create this cultural blend between the observed zones of the city, and inherently creates the final layer of the original Density Diagram in which we re-interpret our analyses into an interstitial negotiation between the culture of Urban Shanghai.

本方案以一种系统的方式对大都市结构内部的活动密度进行研究，并以此作为分析的基础。通过标示场地周围城市文脉中的代表节点，并进一步分析在建成的环境中活动质点之间的相互关系，这个系统将在上海的城市结构内重新建立多个关键区域。基于相邻质点的密度以及控制点分离的程度，这个系统建立了内部的一系列初始点，并且运用一个有机的生长模式，在基地范围内定义了一个基本的空间体量。公共性更强的功能项目的介入将在城市活动区域之间创造一个文化的过渡区域，并最终建立起最初"密度图解"的最后一环。在此图解中我们的一切分析也可以被重新解读为对上海城市文化的一次间接调和。

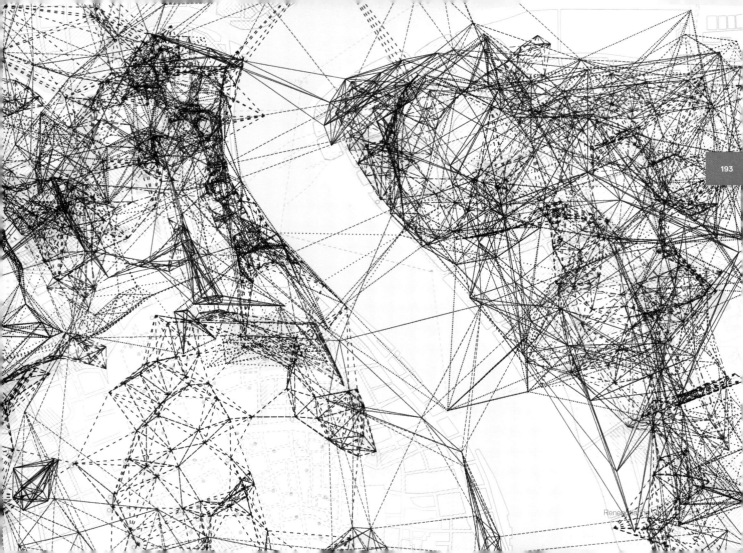

Extracts the 'hidden' densities deep within the fabric of shanghai, as we approach the concept of density as means of analysis rather than a means of evaluation; utilizing a system of adjacencies within the main cultural classes within shanghai, we are able to analyze specific aspects of the city and observe a system of evolutionary sprawl and growth reveals the logic system within which the densities are able to operate as a conceptual driver for the project, as the site will become a 'filter' for flows (information, population) within the general context of the city; in essence, utilizing the site to create an area which promotes interaction between commonly isolated classes within shanghai

提炼上海城市肌理中隐含的密度，我们提出了较评估来讲更接近密度意义的概念，利用一种有着上海主要文化阶层的接近系统，我们能分析城市的特征方面，能观察一个显现了密度逻辑系统的散乱生长进化系统，，这些能作为概念来发展我们的方案，由于本案将成为有着城市基本文脉的流动（信息，人口）的过滤器，由此利用本案去创造一个区域，可以提升上海通常被隔绝的各阶层相互交流。

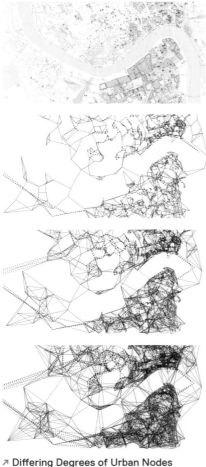

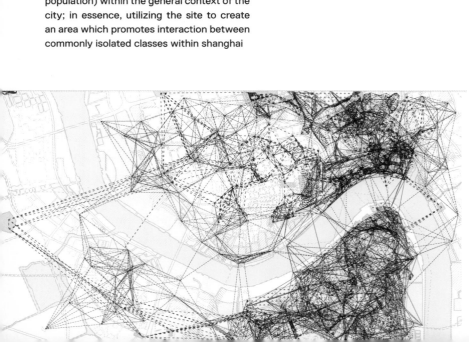

↗ **Differing Degrees of Urban Nodes**
都市节点的不同等级

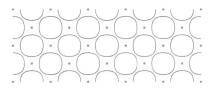
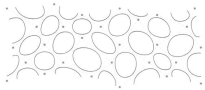

Weaving » Utilizes sight lines as an infrastructure for receiving information and mitigating flow
编织 » 利用看得见的线作为一种能接收信息和缓和流动的城市基础设施

Control Points » Generates a framework on which to build patterns with which to manipulate the weave
控制点 » 生成一种能建立类型和编织操作的框架

Fluctuation » Observes various densities and optimizes the form to respond to all factors
起伏 » 观察各种密度和完善形式以回应所有因素

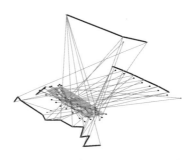  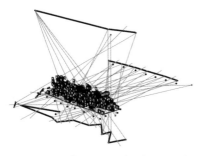

Lower Level » Generates topographical constructs to contain existing activity within the site area
低层》建立地形结构去包含基地已有的活动

Middle Level » Generates the 'sight filter' to describe distribution of main programmatic spaces
中层》建立"视觉过滤器"去描述主要功能空间的分配

Upper Level » Generates visual connections from and through the site towards old town/ pudong
上层》建立从基地到旧城/ 浦东的视觉联系

 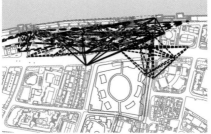 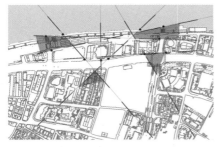

Redefining the Site Filter » By observing its connection between larger activated volumes and sight lines as a means of directing spatial expansion
再限定基地过滤器》通过观察较大活动的体量和空间延展视觉线的联系

Generating Main Spatial Zones » Defines larger densities as main programmatic areas, creates a secondary network of spaces
生成主要空间区》限定较大密度作为主功能区域,创造一个次级空间网络。

Expanding the Site Boundaries » onnects the adjacent open spaces to the upper layers of the space, while keeping the ground plane more free and open
延展基地边界》连接临近公共空间到空间上层,并保持地面更自由和开放

Spatial Density » Generated through interconnections and overlaps observed at sight lines
空间密度 » 在视线观察上建立相互联系和重叠的生成

↙ **Longitudinal Section**
纵剖面

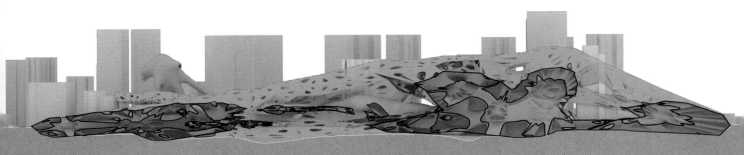

↓ **Private Series**
私密系列

↓ **Semi-Private Series**
半私密系列

↓ **Dendritic Path Series**
树突状路径系列

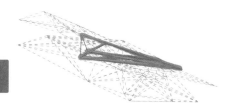

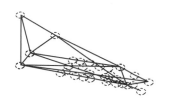

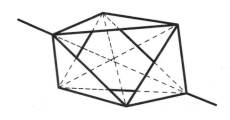
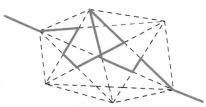
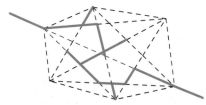

↓ **Interactive Network Series**
相互影响网络系列

↙ **Transverse Section**
横剖面

201

5ᵃ

CLUSTER PROJECTS 团队成果

STUDENTS 学生
Joseph Hines «SoA RPI
Muhan Cui «CAUP Tongji

Decoding the Urban Node
解码城市节点

In a rapidly urbanizing city like Shanghai, immense networks of highways, roads, metro stops and pedestrian thoroughfares are being created to connect people across great distances. Throughout this network exist urban nodes operating at varying intensities and scales. Like that of, Lujiazui, Yu Yuan and Jingan Temple, these nodes in particular are hot spots of local and tourist activity and often contain large scale convergences of infrastructure. This intervention seeks to introduce a series of new urban nodes at various scales within the site. By creating a new hub of transportation and public infrastructure the project aims to attract high levels of activity from both the local and tourist populations. Existing areas of importance surrounding the site are physically connected by large circulation elements. Along these strands exist dense clusters of program which weave through the various levels of circulation using an interlinking operation.

在像上海这样城市化进程快速的大城市中，大量的由高架、道路、地铁站和人行步道等所组成的交通网络也在迅猛发展，使得这个城市中相距遥远的人们之间的联系得以更加紧密。在这个交通网络中存在着不同强度和规模的城市节点。像陆家嘴、豫园、静安寺这些地区，节点是附近居民和旅游者活动的聚集区域，集中了大量的基础设施。本设计试图在基地中引入一系列规模不同的城市新节点。通过建立新的交通枢纽和城市公共设施，吸引大量城市居民和旅游者活动于此。基地与周边现有的重要区域由大型的交通流线元素直接连接起来。密集的功能区域沿这些线性元素布置，与不同层高的交通流线交织在一起。

Decoding the Urban Node

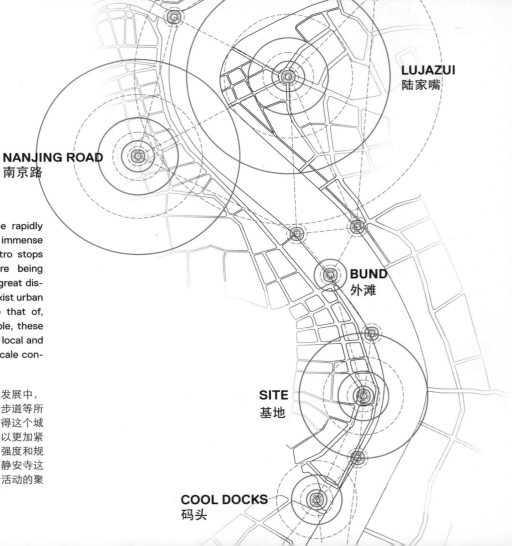

Site Map and Analysis » In the rapidly urbanizing city of Shanghai China immense networks of highways, roads, metro stops and pedestrian thoroughfares are being created to connect people across great distances. Throughout this network exist urban nodes of varying intensities. Like that of, Lujiazui, Yu Yuan and Jingan Temple, these nodes in particular are hotspots of local and tourist activity and contain large scale convergencies of infrastructure.

基地分析 » 上海在城市化进程快速发展中，大量的高架、道路、地铁站和人行步道等所组成的交通网络也在迅猛发展，使得这个城市中地处遥远的人们之间的联系得以更加紧密。在这个交通网络中存在着不同强度和规模的城市节点。像陆家嘴、豫园、静安寺这些地区，节点是附近居民和旅游者活动的聚集区域，并集中了大量的基础设施

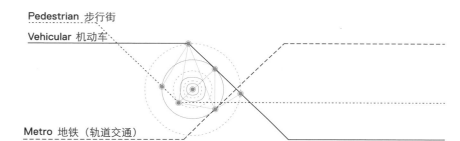

Pedestrian 步行街
Vehicular 机动车
Metro 地铁（轨道交通）

↗ Informal Food Vendor Markets
食品自由市场

↗ Organized Gatherings (taiji)
健身活动

↗ Organized Gatherings (dancing)
文娱活动

↗ High Commercial Activity (international)
高密度国际化商业街

↗ High Commercial Activity (local)
高密度本地化商业街

↗ High Commercial Activity (informal)
高密度商业自由市场

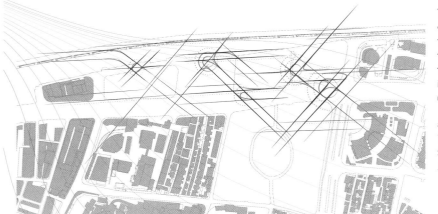

The site in its existing condition lacks the conditions assoiated with the notion of the urban node. Despite its close proximity to the Yu Yuan, Cool Docks and The Bund, there is nearly no activity taking place. This intervention seeks to establish a series of uban nodes within the site at various scales

基地在现有条件下缺乏与城市节点的互动。尽管其靠近豫园、老码头和江边，但几乎没有活动的地方。这种干预的目的是企图在基地中建立一系列不同规模的城市性节点。

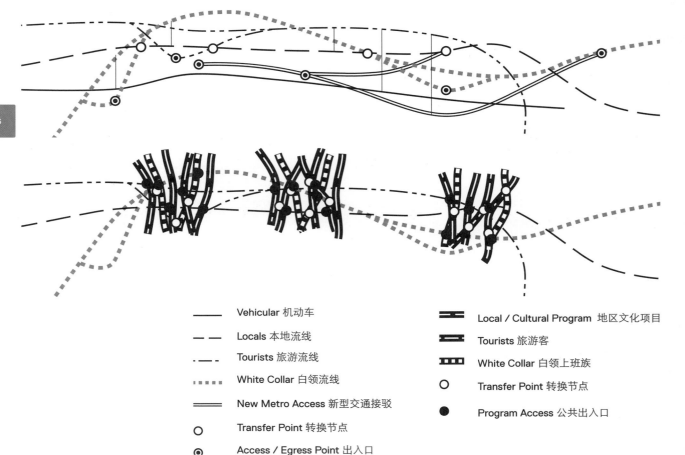

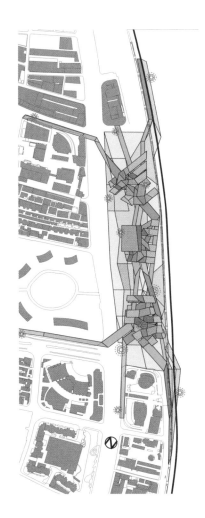

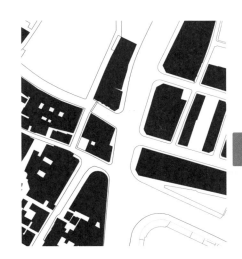

When looking the site plan it is clear that there are three dense regions. These regions house the program elements and exemplify a complexity similar to that of the old city. The circulation strands extend to the city in an effort to reach key points to being to draw people into the architecture.

基地周边很明显由三个密集的区域形成，这些区域的建筑元素体现老城区类似的复杂性，并循环延展到城市，引入建筑学的意义。

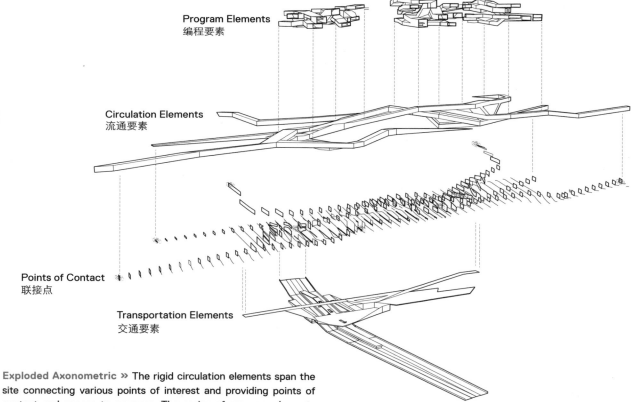

Exploded Axonometric » The rigid circulation elements span the site connecting various points of interest and providing points of contact and access to program. The series of program elements weave themselves throughout the rigid. The interstitial spaces within this assembly provide semi-enclosed spaces to be used for cultural gatherings and installation.

分解轴测关系 » 基地的各类活动单元通过刚性流通的空间连接，并形成相关系统。系列的元素以符合自身需求的次序编织而成，在此建构体错落而成的半开放空间提供了城市文化交流活动场所。

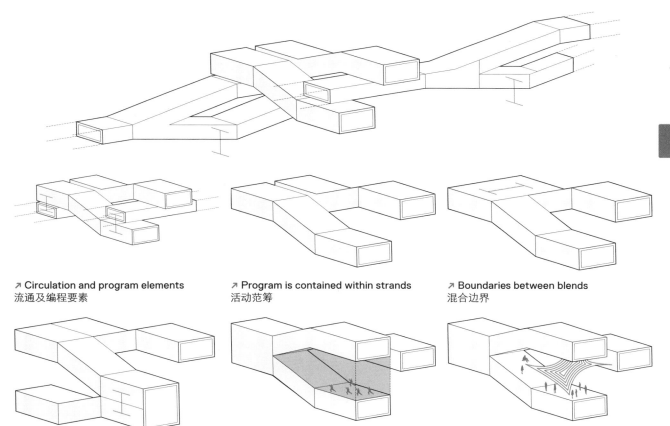

↗ **Circulation and program elements**
流通及编程要素

↗ **Program is contained within strands**
活动范筹

↗ **Boundaries between blends**
混合边界

↗ **Vertical blending creates opportunity for larger programs**
竖向空间叠合提供大型活动空间

↗ **Indoor/outdoor spaces are created to hose local cultural programs**
管状空间交织提供了室内外地区文化活动空间

↗ **Insitutional spaces are used for large scale art insallations**
场景空间作用于大规模的艺术活动

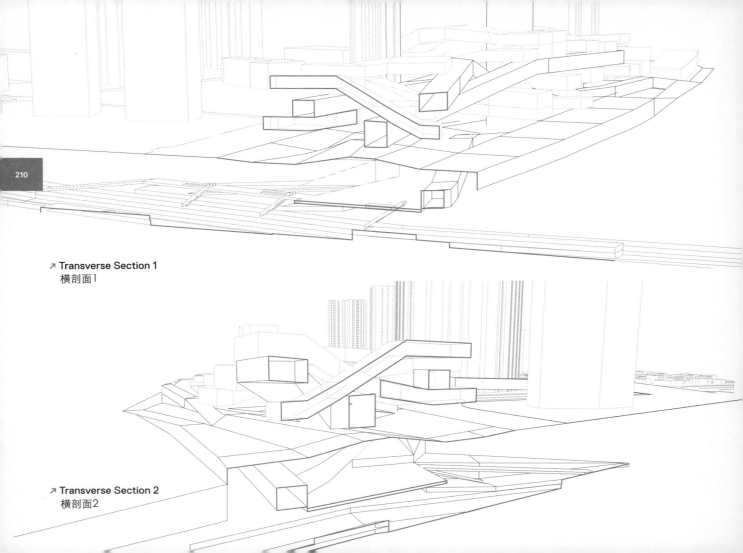

↗ Transverse Section 1
橫剖面1

↗ Transverse Section 2
橫剖面2

5^b

CLUSTER PROJECTS 团队成果

STUDENTS 学生
Dana Shin «SoA RPI
Aristan Yudhi «CAUP Tongji

Layer, Interrupt and Connect
分层，中止，连接

Waitan, located in the heart of Shanghai, attracts and redistributes the major flows of movement in the city. Our site, Shi liu Pu, is influenced both by the Bund, Pudong area and the South station, yet not active enough to attract people in as the fabric itself. Our proposal was to redevelop this area, located right below Waitan to redirect the movements of pedestrians and maximize the benefit of different modes of transportations on the site. The relationships among the surrounding areas such as Pudong, The Bund, Xintiandi, Yu Garden, Old City, Cook Docks, and the South station create woven connections that start to inform us an overall behavior of the site. Our approach was to inherit these gestures and apply to reshape in such a way it can bring people in not only as means of transportations but also as public spaces. By exploring the potentials of interworking through physical models of tectonic patterns, we inherited structural factors of interweaving; layer, interrupt, and connect.

位于上海市心脏地带的外滩是城市交通的重要集散地。十六铺基地为外滩、浦东开发区和一个长途车站所包围，但仍不能有效吸引城市人流。本方案希望从新发展外滩南部这一地区，从新引导人流，并使周边各种交通方式最优化。基地与周边的浦东、外滩、新天地、豫园、上海老街、老码头和汽车站形成一系列交织在一起的连接关系，这些连接反应了场地的整体情况。本方案继承了城市文脉中的手法，从新塑形，在建立交通枢纽的同时建立公共空间，吸引公众进入场地。在以实物模型为基础、探索编织构造的各种可能性的过程中，编织的结构特征：分层、中止和连接被继承下来，运用到建筑设计之中。

Layer, Interrupt and Connect

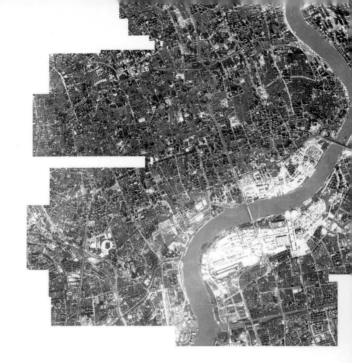

浦东
Pudong

老码头
Shanghai Docks

现存建筑
Existing Building

居住社区
Residential Area

Movements in microscale introducing woven fabric of the site.
基地表征的微观动态

Existing lines of traffic.
交通现状流线

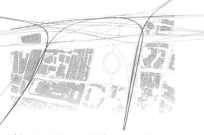

Linear gestures of the secondary connectors.
辅助的线性规划链接

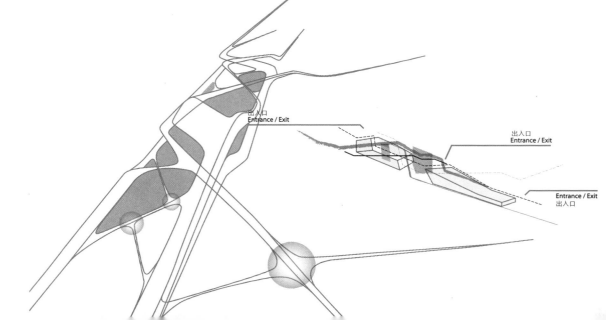

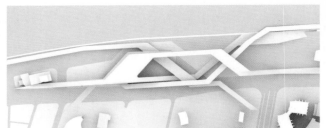

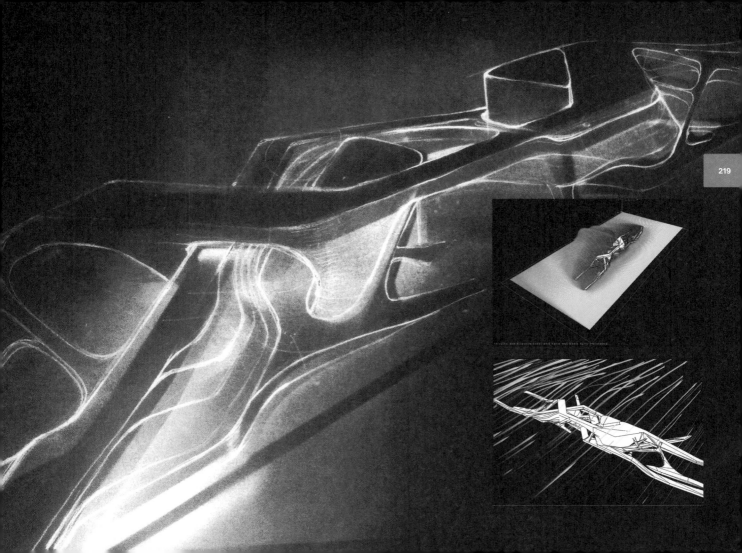

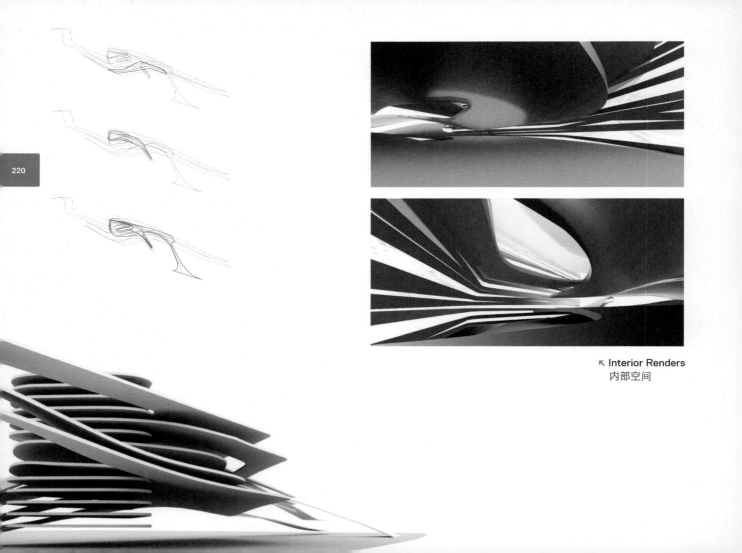

↖ **Interior Renders**
内部空间

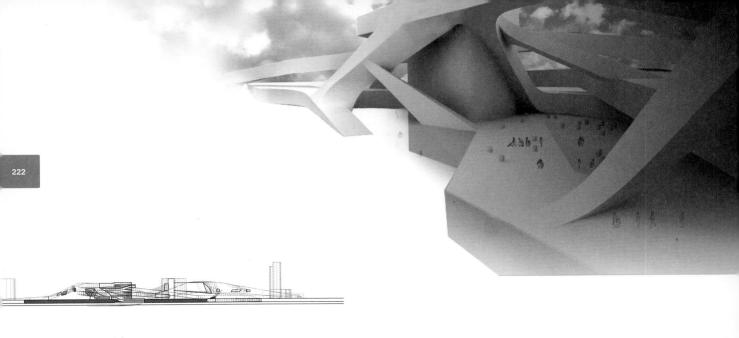

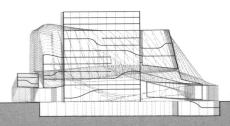

↗ Longitudinal Section
纵剖面

↗ Transverse Sections
横剖面

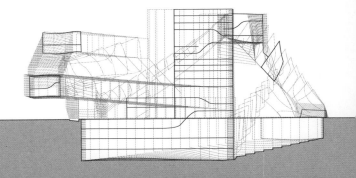

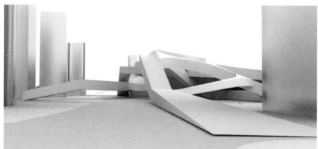

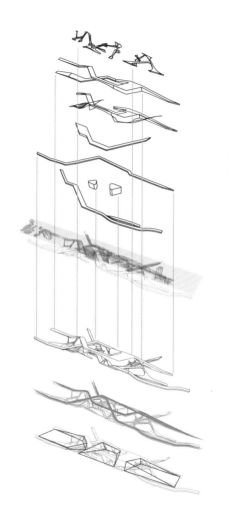

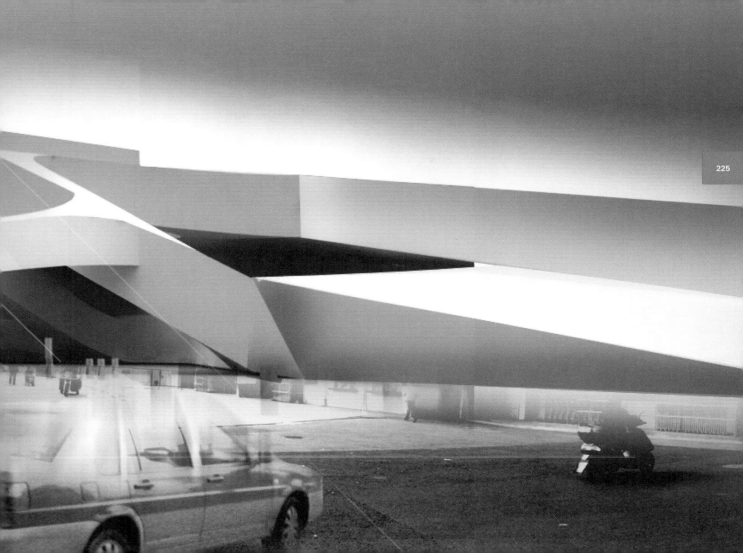

6ᵃ

STUDENTS 学生
Jamie Lee «SoA RPI»
Bowen Zhang «CAUP Tongji»

Bifurcating Urban Contours
分支城市轮廓

The drive of the urban fabric is one that is defined by the movement of pedestrians, automobiles, subways and buses, all weaved together to form the city of Shanghai. Mapping these methods of movement help to reveal situations in which people conform and disperse throughout the city, bifurcating the solitary line that is thought to be traveled. This can be seen in a vertical distribution as well since transportation, whether on foot or by subway, exists in a three-dimensional field rather than a two-dimensional one. The curves generated by the movement of the urban cityscape helped to influence the nature of the structures on site above and below ground. The weaving context was reintroduced as the forms of our buildings motivated by the necessity of the site to provide for an interesting and alluring spatial condition both outside and inside. We introduced contours as a way to deal with the emergence of the ground on site as well as to provide a façade that allowed for an environmental continuity with the city.

城市肌理在行人、车辆、地铁、公交车等诸多元素的流动中不断更新。这些人与车的流线相互编织，形成了上海复杂多变的城市特征。在对不同流线的描绘中，城市中直观感知的路线被不断地分支和重组，构成复杂的功能关系网。这种抽象网络便将人们在城市中汇聚和分散的行为详细地揭示出来。在垂直上的分布中，这种不同高度的活动之间的联系依然存在，因为不论是地铁还是人行流线，都存在于三维交织的而并非平面的场域中。在城市空间运动的研究中建立的曲线网络影响着基地地上和地下的空间结构。在建筑形态的生成过程中，编织的概念被再次运用。这种对具体编织形式的运用是缘于基地现状需求一种有趣而有吸引力的内外空间形式。同时，在场地设计中引入了起伏的城市景观以建立一种城市中具有环境连续性的表皮。

Circulation and Spatial Distribution » The rigid circulation elements span the site connecting various points of interest and providing points of contact and access to program. The series of program elements weave themselves throughout the rigid. The interstitial spaces within this assembly provide semi-enclosed spaces to be used for cultural gatherings and installations

循环及空间分配 » 基地的各类活动单元通过刚性流通的空间连接,并形成相关系统。系列的元素以符合自身需求的次序编织而成,在此建构体错落而成的半开放空间提供了城市文化交流活动场所。

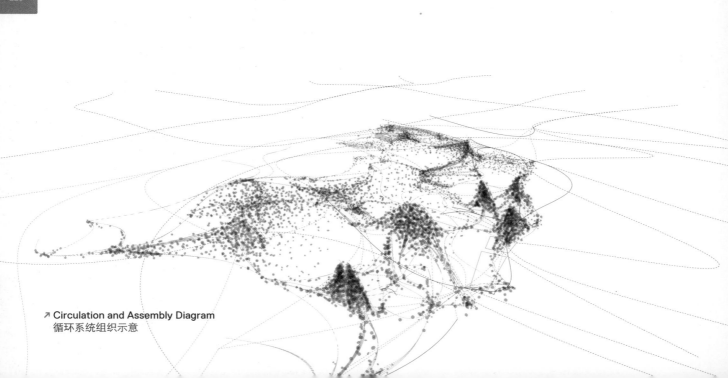

↗ Circulation and Assembly Diagram
循环系统组织示意

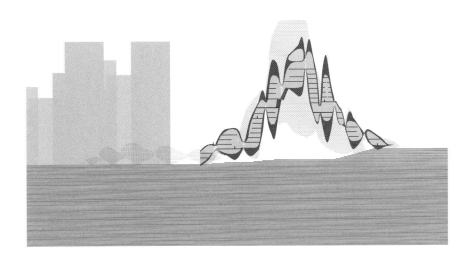
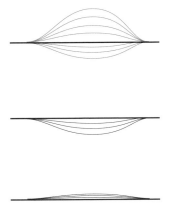

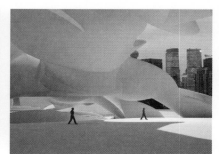
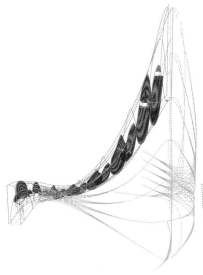
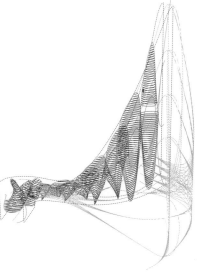
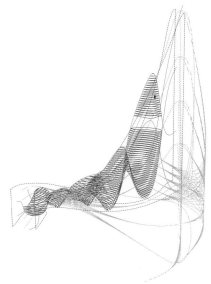

232

↗ Various Plan Cuts
分层平面

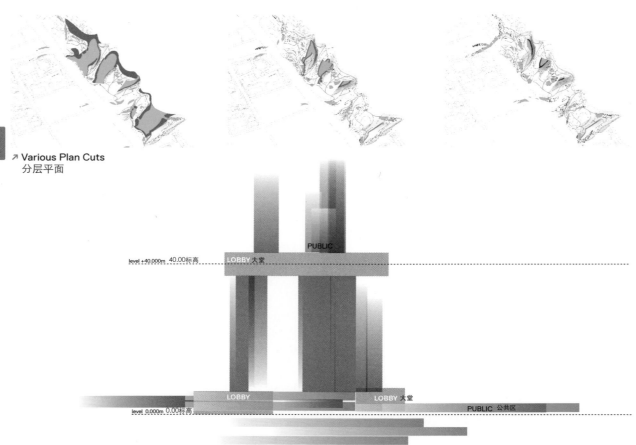

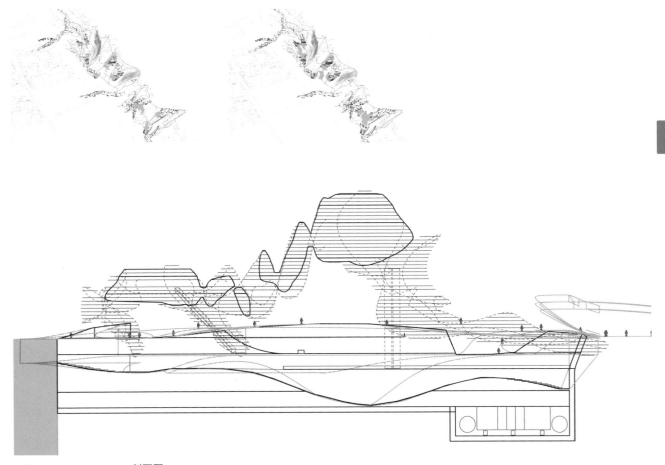

↗ Diagrammatic Section 剖面图

235

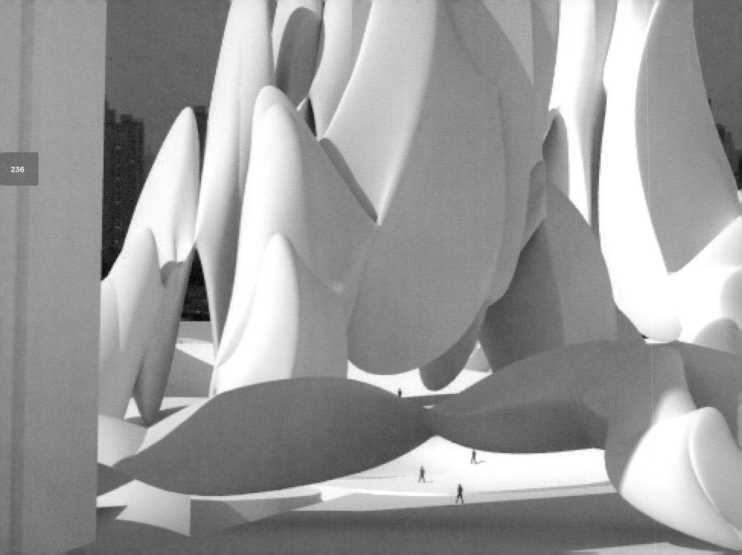

6ᵇ

CLUSTER PROJECTS 团队成果

STUDENTS 学生
Justin Rupp «SoA RPI»
Hani Shin «CAUP Tongji»

Linked Alterations
连结的改建

The focus of linked alterations is to monopolize on the versatility of woven fabrics and systems. Experiments in weaving were fundamental to the project and the goal of these experiments was to be able to fully understand the linked weaving system. By understanding the structure and accurately understanding how the links work to generate a surface and how the weave works as a surface, one can begin to more easily manipulate the geometries. A system of towers and roots covers the landscape along the South Bund, pulling people into and moving them through the site. Literal and figurative weaving are key elements on all scales of the project; connecting the surrounding city and HuangPu River, strategically interconnecting the programs on the site and developing a variable woven façade system. By pulling these parts of the city, river, and site together in a single woven landscape, softer edges are created along the site.

本方案致力于实现编织结构系统多功能性的最大化。进行编织的探索性实验对本方案格外重要，其目的便是充分理解编织系统。通过研究结的结构、具体掌握如何利用结手法生成面以及如何在面的基础上编织，对类似几何体的操作变得更为简单清晰。在具体实践中，一系列的塔和根覆盖了南外滩边的景观区域，引导着人们进入并穿过场地。具象的编织形态是整个项目中各种尺度下的关键元素。这些形态连接着周边城市和黄浦江，有意将场地内的项目功能连接在一起并发展一套富于变化的表皮系统。通过将城市、江水和场地集合为一个编织的城市景观，场地的边缘可以与周围环境更好地融合。

The forms are generated through a series of steps that question the regular arrangement of buildings. Atriums, high rises, and the ground condition are all re-evaluated in order to maximize an interaction between distinct groups of people.

An atrium is a central hall or open courtyard in a building. The idea is to increase visual interactions within the building. However, our goal is to expand the area in which people are able to interact whether it be physically or visually. For example, by shifting the core to the exterior of any given floor the views instantly begin to expand out. Instead of only viewing people within a single building, an individual may begin to view into surrounding buildings therefore creating visual connections where there were none before and thereby creating an exchange.

(中文注解见P242)

↗ Viewports and Tower Elements
景观视点及塔楼要素分析

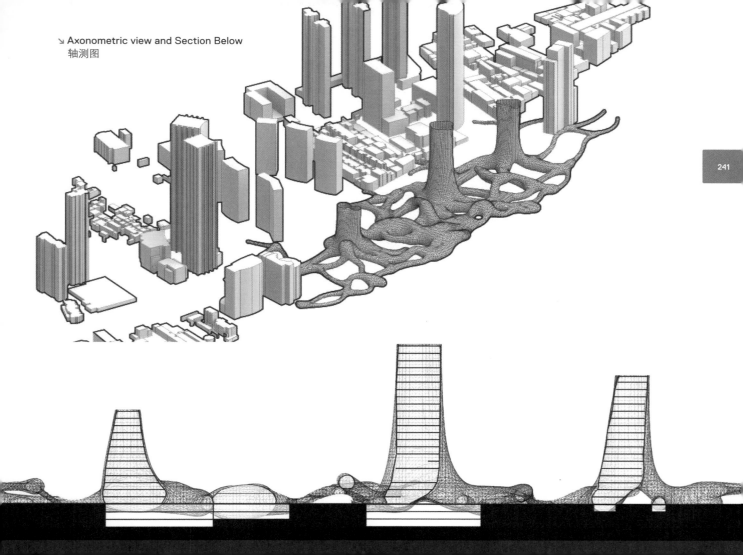
↘ Axonometric view and Section Below
轴测图

通过一系列手段并面对各种建筑构成条件而生成的形式，最大限度地提高不同群体的人之间的互动。

正如建筑物中的中庭或开放性院落，其目的是提高在建筑物内的视觉交互。而本案的创作目标是扩展这种交互影响，包括物理的或视觉的作用。比如，通过任何给定的视点，从内部拓展到外部空间，取而代之的是观察者可以从一个独立的建筑视角转化为对建筑群组的空间认知，从而创造前所未有的视觉连续性。

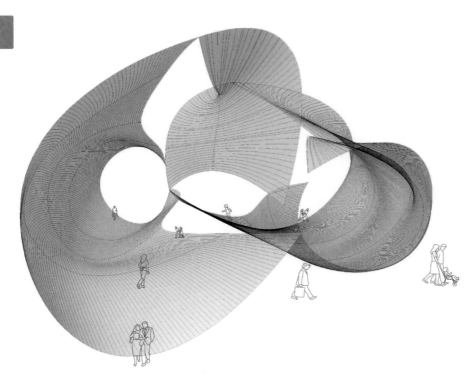

CLUSTER PROJECTS 团队成果

7ᵃ

STUDENTS 学生
Anthony Policastro «SoA RPI
Zimei Shen «CAUP Tongji

Meta-Desic Architecture
单元架构

From initial studies, it was determined that the site had the potential to become a hub of transportation and a cross roads for both tourists and locals. As a result, various programs had to be accommodated for and arranged in a logical manner. As the project progressed the form began to grow from concepts of packing and braiding, thus this mode of production became a nesting style of working. To start, a super structure was developed to house substructures. Next, the substructures were placed using various calculations and data. To bring the project full circle, both forms underwent a tectonic development to further understand how their skins and construction logics could potentially function and develop at a more humanistic and relatable scales. The program on the site was determined most appropriate for the site, in hopes it would begin to function as a beacon of culture, business, and public interaction.

经过初步研究，基地区域有望发展成为游客和本地人的交通枢纽。因而不同的功能必须按照一定逻辑布置其中。包裹和编织的概念构成了形态的生长逻辑，这种设计模式循环嵌套，逐步深化。首先在基地上建立巨型尺度结构，并将子结构设在其中。然后，通过各种计算和数据，明确子结构。为贴近使用者的尺度、使表皮和建构逻辑更加合理，我们进一步深化了巨型尺度结构和子结构的具体构造。本方案的功能项目均依照场地的需求安排，希望能发展为一个文化、商业和公众交流的灯塔，吸引游客和本地人。

Meta-Desic Architecture 测元线架构

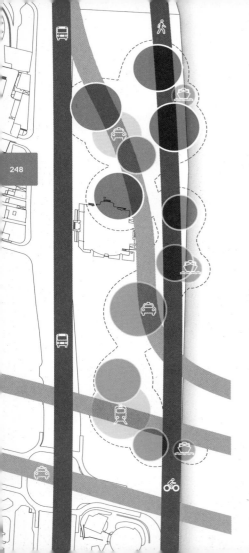

After visiting the site, we came to realize that much needed to be done to enliven the area and connect it to the city and various tourist and Shanghai landmarks. Our initial site diagrams focused on understanding the sites various narratives and speeds. This it to say how a user, tourist or local would begin to use the site and at what rates the user would move thru and from the site. By working in such a manner we began to carry Shanghai through the site and connect the site as a hub.

As a foreigner we began to visit various tourist sites within the city to understand the flows and underlying circulation strategies already within Shanghai's fabric that were catering to foreigners. The findings were that in many cases, private buses, taxi cabs, or the subway lines were the easiest and most used modes of transportation.

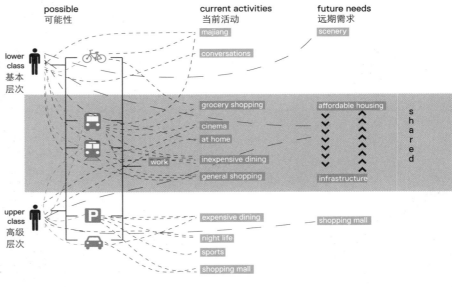

根据基地情况，我们需要提升区域活力，并连接城市各种旅游活动成为上海的地标。从最初的现场图表集中了解基地的各种情况，了解本地居民和旅游者在基地内活动的比率，并通过这样的工作方式，建立一个连接上海城市活动的枢纽。

研究发现在许多情况下，公交、出租车或地铁线是最简单和最常用的交通方式。

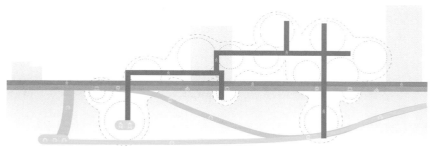

↗ Diagrammatic section of transportation
竖向交通衔接

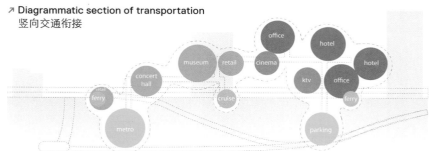

↗ Diagrammatic section of program
竖向功能分布

通过对区域的旅游、文化等方面的调查研究，我们定义该地区人口密度和办公、酒店为主的业态分布，以满足在未来的城市品质。因此，我们建议的区域功能重点以娱乐、商业、零售、写字楼、酒店空间、公共空间，以及交通枢纽为主。

To understand the program we researched Chinese tourism, culture, and the surrounding area. The findings concluded a program that revolved around middle to upper class Chinese users with a small portion allotted to foreign tourists. After researching the area, we determined that the area was heavily populated with office spaces and hotels. In addition, the area is projected to cater toward the middle to upper classes in the future. As a result, our proposal catered to the middle to upper class with a heavy focus on entertainment, commercial, retail, offices, hotel spaces, public space, and major transportation hubs.

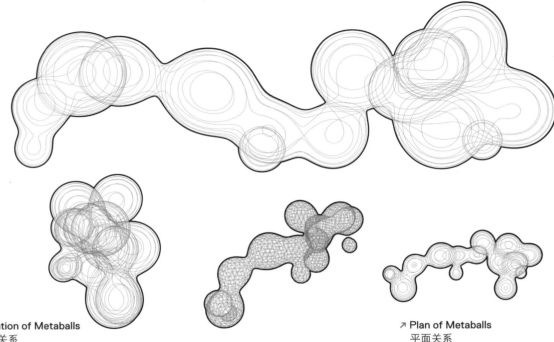

↗ Elevation of Metaballs
竖向关系

↗ Plan of Metaballs
平面关系

As an initial experiment, Grasshopper and excel were used as exploratory tools to understand what and how the program massing could be arranged across the site. Excel was used to arrange and calculate basic square footages, daylight, public space, and unit numbers. With this information calculated and arranged, a grasshopper script was used to produce meatballs from predetermined nodes within the site. As a result a certain packing logic began to appear. In addition, the experiments began to bring the project full circle and inform as to where and how circulation and public space would function within the context of the buildings. The points themselves developed as the program spaces, whereas circulation and public space was allotted to the voids. Using these excels sheets and grasshopper scripts we then began go into the units themselves and develop them in the same manner, much like a Russian nesting doll.

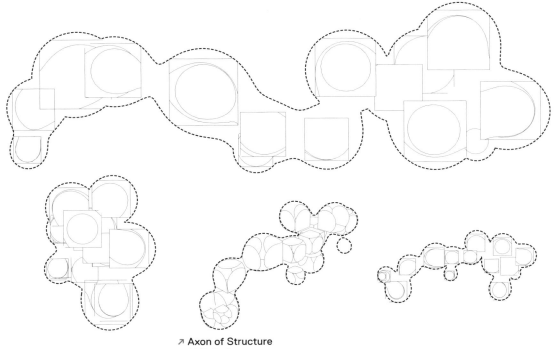

↗ **Axon of Structure**
结构关系

根据这些信息计算和编制，Grasshopper软件被用来编制生成基地内预设的功能气泡节点关系，形成相应的逻辑。此外，实验开始完整地把项目引入程序，设定建筑物的公共空间以及其作用范围。运用这些软件的优势，使单个单元以同样的方式循环生长，如同一个俄罗斯嵌套娃娃。

The building was understood as two towers positioned at both ends of the site that reached towards one another and met at the middle of the site in front of the pre-existing Bank. Tower One, to the north of the site, was anchored by a subterranean parking structure that connected the newly proposed transit tunnel system. Here, locals could park their cars or be dropped off by public transportation and then travel up into the builing. As you progress up through the tower you would move through office space, commercial space, and a hotel at the top.

该建筑被理解为由基地两端的塔楼与基地内保留的银行大楼组成。北侧塔楼链接一个地下停车场并与过境隧道系统相接，这里有良好的可达性。穿越其中，关联了办公空间、商业空间以及顶部的酒店。

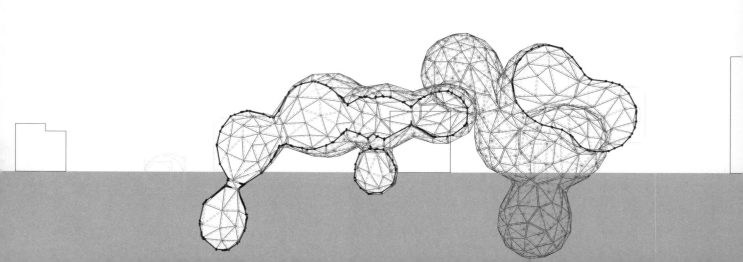

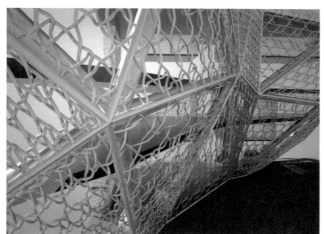

7b

STUDENTS 学生
Michael Mancuso «SoA RPI»
Yinjia Gong «CAUP Tongji»

Spatial Void Inter-looping
空间内循环

A vast site surrounded on all sides by a variety of programs oriented towards people of a variety of backgrounds and income levels necessitated an intervention that could both divide the entirety into manageable human-scale volumes while also maintaining both visual and physical connections throughout. The volume of the site then becomes a solid container with an intervention of braided voids that serve to divide the volume and allow for programmatic volumes. This simultaneously provides for circulation throughout the site, as well as visual connectivity from the street to Pudong. The volume of the site is a negotiation with a variety of forces inherent in the site. The roof negotiates the Indigo Tower (north), the Commerce Center (south), and the Huanpu River (east). The ground negotiates access to the site from the street, the Bund, the river, the Cool Docks, and the Expo site. A horizontal weaving of void space occurs through the volume, defined by the circulation. This integrates and blurs the interior and exterior spaces, opening up the massive site physically and visually.

这个基地的特殊性在于其周边有各种不同的功能配置，所服务人群的背景和收入情况也很复杂。在这样复杂的情况下需要介入一种场所既要将整体划分至人的尺度又可以保持视觉上和实体上的联通。场地的体块被一系列功能体量分割，进而建立了场地上的交通流线，同时还保留了街道到浦东的视觉通道。这个体量是与基地周边不同因素对话的结果：屋顶造型形成了与酒店（北部）、贸易中心（南部）和黄浦江（东部）的对话，而地面形成了与街道、外滩、黄浦江、老码头和世博园的对话。一个受交通流线影响的水平空间穿插在整体体量之间，连接并模糊了室内外的空间，在视觉上和空间上打破了场地中的阻隔。

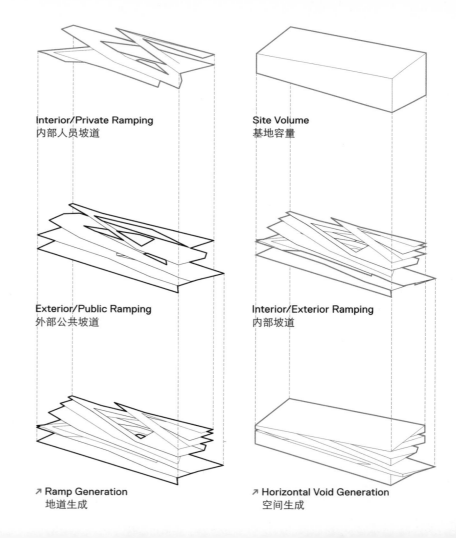

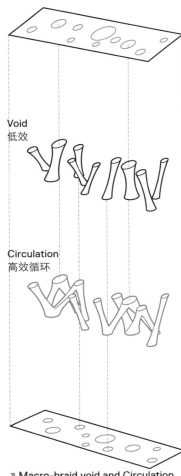

Void
低效

Circulation
高效循环

↗ Macro-braid void and Circulation
客观编织空间循环

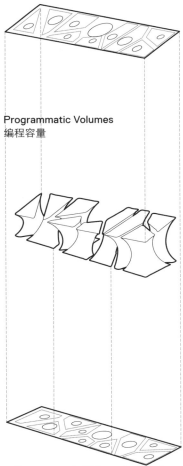

Programmatic Volumes
编程容量

↗ Programmatic Volume Generation
编程容量生成

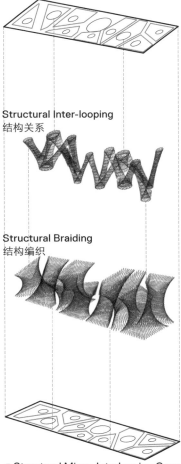

Structural Inter-looping
结构关系

Structural Braiding
结构编织

↗ Structural Micro-Interlooping Generation
微观结构系统生成

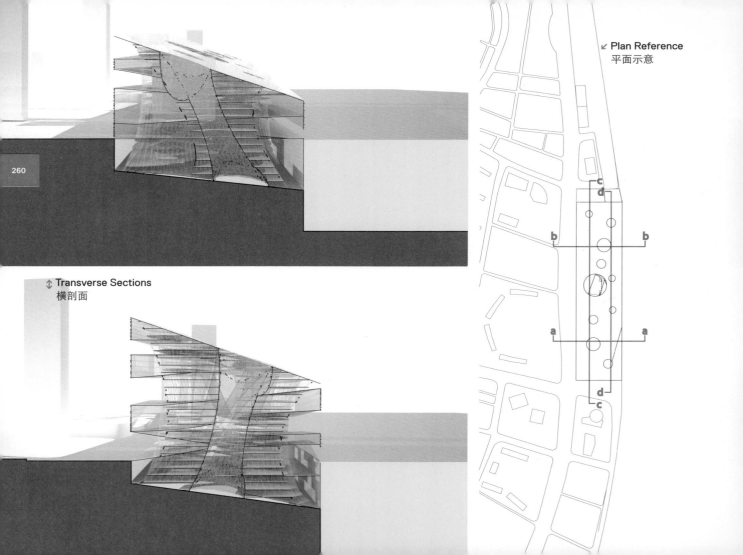

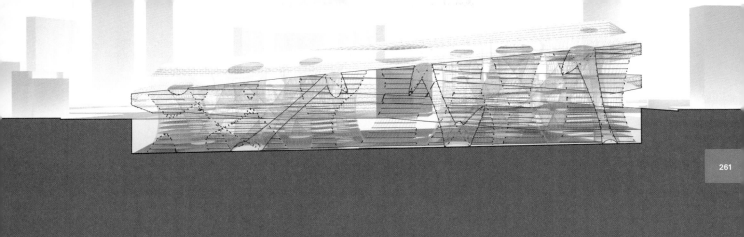

Longitudinal Sections
纵剖面

8ª

STUDENTS 学生
Caitlin McCabe «SoA RPI
Zie Jie «CAUP Tongji

Fluid Waves
流体波

Our site is along the Huang Pu River in Shanghai, China. This site When starting this project, we first studied the different types of people that interacted with our site and their different modes of transportation getting to and from our site. We also looked at the surrounding streets that feed into our site. One thing that interested us the most was the changing of small to large scale program along Old Shanghai Street, from Yu Yuan leading towards our site. We took that idea of small and large scale program and brought back our early studies on weaving. So we started to look at the different scenarios of how to interconnect these two types of spaces and see them as one instead of seeing them as drastically different. From this idea, we got this wave-like project: a mix of landscape and architecture, where you can be on the ground floor of one building and walk a few steps and be on the top of another. There is a blurring of what is exterior and what is interior.

方案初期我们分析了位于上海黄浦江边基地之上不同交通工具的不同流线，研究它们如何进入基地、人们如何与基地互动。同时我们也关注了从豫园到基地一路的城市肌理。其中最值得引起关注的是如何处理小尺度与大尺度建筑之间的差异。通过将两种尺度的概念结合到早期的编织研究中，我们试图在它们之间建立连接，考虑其中的共性而非差异。最终这个混合了景观与建筑的波形体就是基于这样的理念诞生的。一部分的地坪在几步之外就是另一部分的屋顶，室外空间与室内空间的界限也因而模糊。

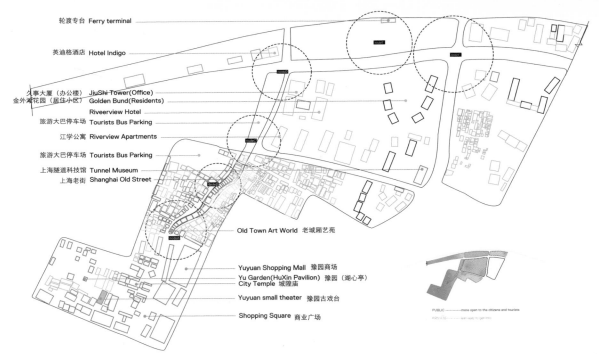

When first looking at the site you start to see the different types of circulation that happens through our site based on the existing surroundings. There are the tourists and the residents all of different classes that travel through our site. Those different pedestrian circulation start to overlap and interweave with one another. YuYuan is a popular public space for both tourist and residents alike. The path from YuYuan to our site is comprised of very dense, small scaled areas and then starts to move to larger scaled areas. Our concept focuses on intertwinging these small, densely pact areas and these large, open areas to allow a different type of feeling you get when coming from YuYuan to our site.

基于基地现存的周边环境，你可以看到不同类型的建筑风格的循环。有游客和本地居民穿越基地，这些不同的路径彼此重叠而交织。豫园是一个受欢迎的公共空间并为不同的人群所接收，连接豫园到我们基地的路径是由非常密集的小型社区，过渡到较大规模的地区。本案的设计理念是通过从豫园到基地的路径关联，混合串联并编织这些小而密集的区域以及这些大型而开放的领域，形成丰富的空间感觉。

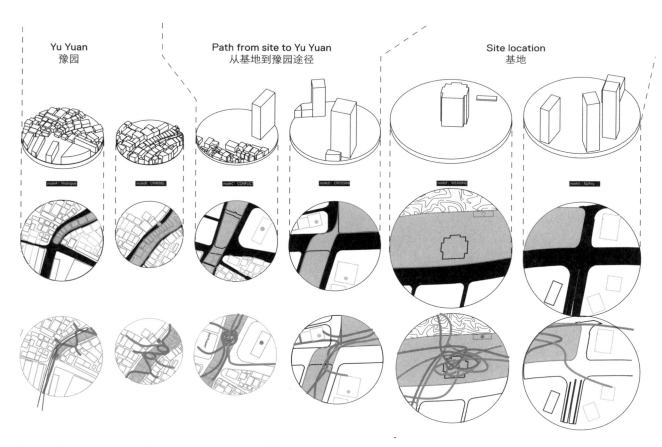

Circulation Diagrams » Small and large scale building types intertwining together bringing all types of people to the site.
系统图式》大小建筑规模和各种类型的综合编织引导各种人群进入基地

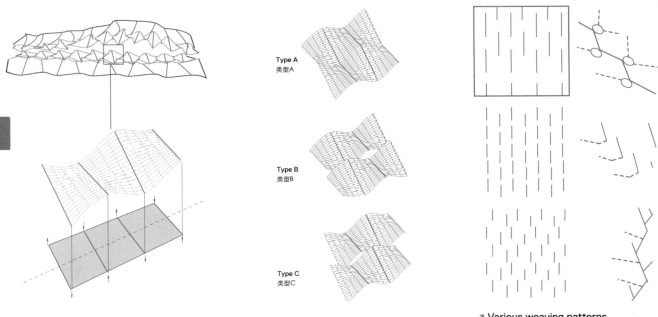

Type A
类型A

Type B
类型B

Type C
类型C

↗ Various weaving patterns
各种编织图式

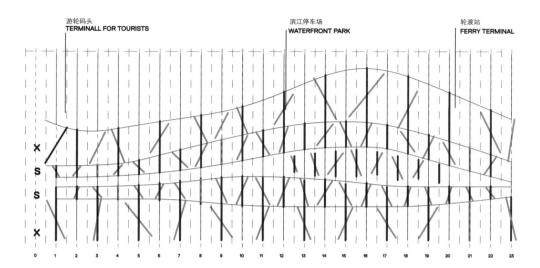

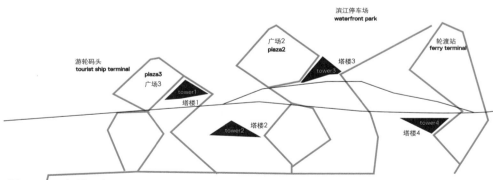

↗ Diagrammatic Siteplan
基地平面图表

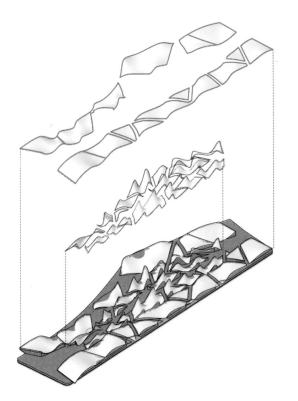
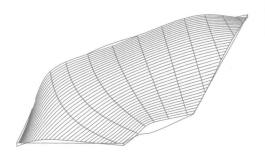

↗ Large scale landscapes with programs under the artificial ground
大型人工地面景观

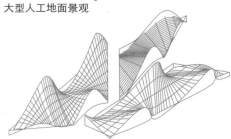

↗ Small scale of shops and programs above the artificial ground.
人工地面上方的小型商业体

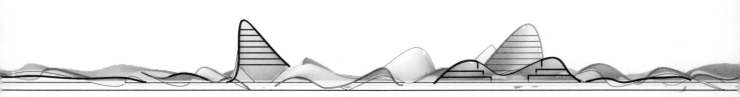

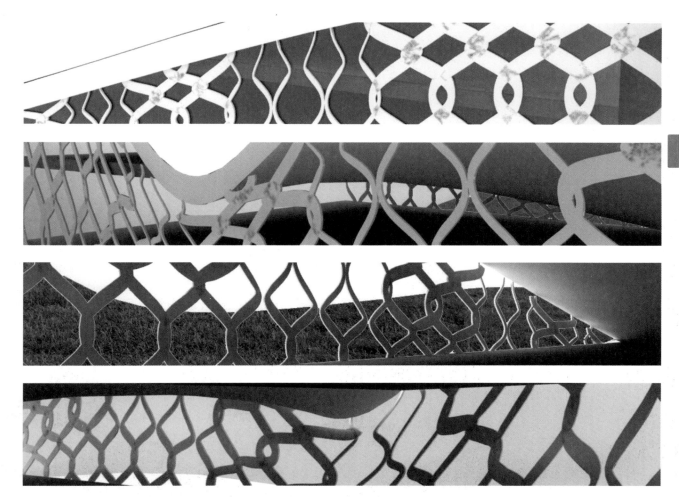

8^b

CLUSTER PROJECTS 团队成果

STUDENTS 学生
Arthur Adams III «SoA RPI
Ning Wang «CAUP Tongji

Shrouded Connections
覆盖连接

The intention of the design was not to simply create a layered structure like most of Shanghai has been planned and quickly developed. Rather we strived to create a 3-dimensional design through a series of weaved spaces for various scales of program. We focused mainly on analyzing the Bund area, but rather than simply analyze why people were attracted to this location, we thought it was more prevalent to analyze the pathways which led people to and away from the Bund. Through this analysis we were able to classify the various streets into four main categories. We used these categories as literal "streets" to weave space vertically and horizontally through our site. Based on various control points we developed which both surrounded and were within the site, we created the large scale locations using 3d voronoi definitions. Then we applied the same techniques to create the small scale pathways which were to connect the various large spaces into this organization system with a structural minimal surface.

方案初衷不是简单地设计一个像上海大部分地区已规划或建造的层状结构建筑。相反，我们试图通过一系列编织三维空间的方法为各种尺度的活动项目建立发生的场所。因而不仅要分析人群为何被吸引到这里，还要研究外滩的步行通道如何引导人流。基于这些分析,城市街道可以被分为四种主要类别，这四种街道在基地内部被垂直地与水平地编织起来。基于前期选定的基地周边和基地上的控制点，我们通过3d voronoi的技术生成大尺度几何体和小尺度的连接通道，最终覆盖以有支撑结构的极小曲面。

Shrouded Connections

Our site analysis began with an analysis of Shanghai from the bund into the city. We organized each main street into one of four main categories. We then took these new "streets" and literally wove them through our site. We chose the pathways that these streets would follow by analyzing various factors around the site such as the locations of businesses and local stores, residential areas, tourist areas, and various transportation locations.

As a local, we began to pinpoint landmarks that were off the beaten path, but as important as the city landmarks. Again, we began to understand the city through the circulation and transportation. From these studies we found that the various bus lines, metro lines, and private bikes were the predominant modes of transportation.

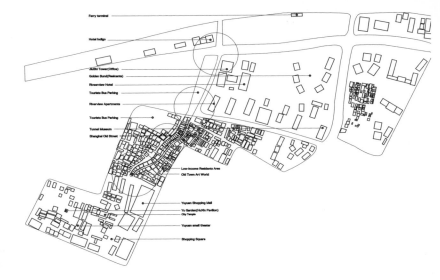

从分析开始，了解外滩地区的上海城市定位。方案组织了主要街道的四大类型，并把这些新定义的"街道"编织在基地之内。我们选择的途径是通过分析基地周边的各种城市因素，如当地的企业和商业、居住区、旅游区，以及各种交通设施的位置，以确定这些街道的相互关系。

这种错纵的街道形式犹如城市地域的标志一样重要。另一方面，通过研究发现，各种公交线、地铁线和自行车也是本区域最主要的交通方式。

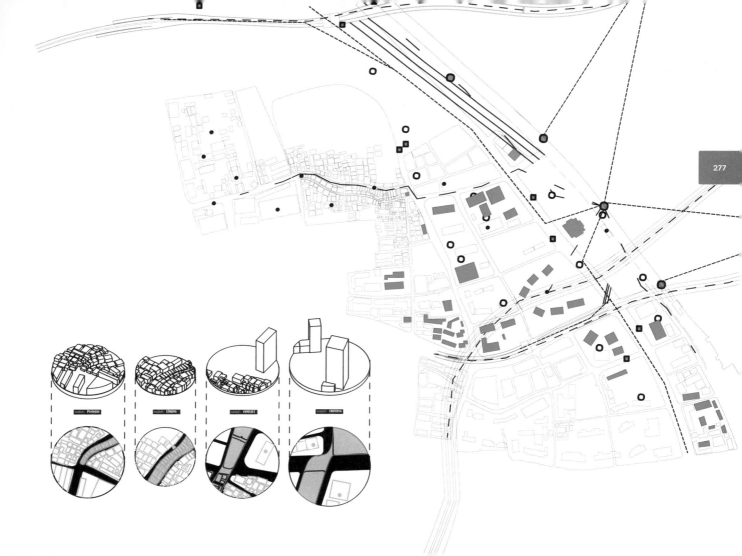

 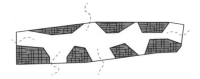 

↗ **Planar Circulation**
平面系统

  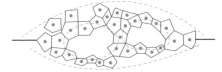

↗ **Sectional Circulation**
剖面系统

  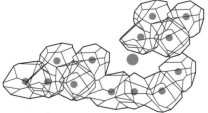

↗ **Voronoi Circulation**
空间系统

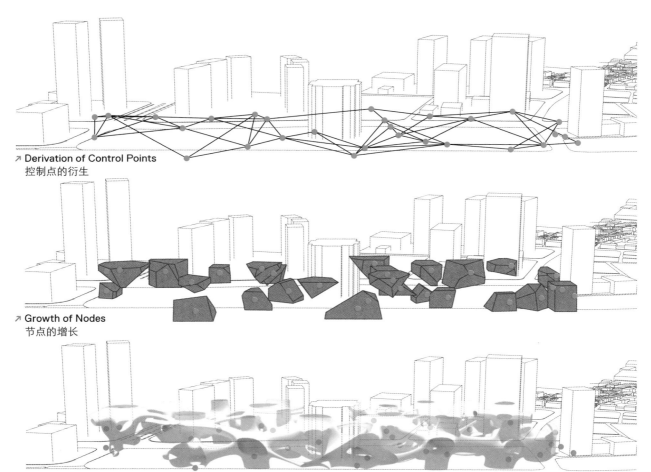

↗ Derivation of Control Points
控制点的衍生

↗ Growth of Nodes
节点的增长

↗ Systematic Connection
系统连接

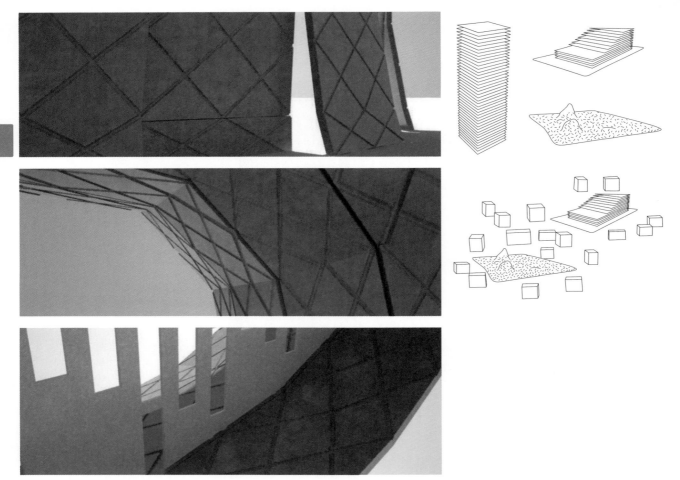

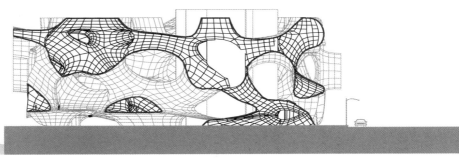

↗ Transverse Section 1
横剖面1

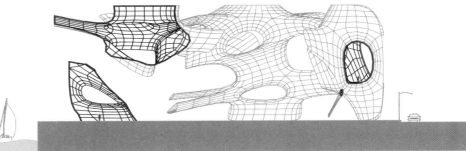

↗ Transverse Section 2
横剖面2

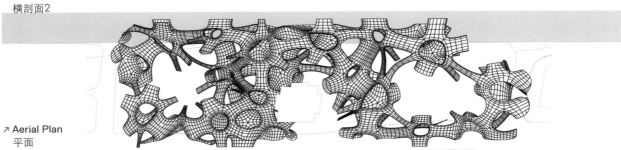

↗ Aerial Plan
平面

Studio Culture 设计文化

RENSSELAER ARCHITECTURE in collaboration with

TONGJI UNIVERSITY

图书在版编目(CIP)数据

交织　上海南外滩地块城市设计：汉英对照/陈宏，魏崴，(美)克伦贝尔(Gustavo Crembil)编著．—北京：中国建筑工业出版社，2013.12
ISBN 978-7-112-16071-6

I. ①交… II. ①陈…②魏…③克… III. ①城市规划－建筑设计－作品集－上海市－现代 IV. ①TU984.251

中国版本图书馆CIP数据核字（2013）第263486号

责任编辑：徐　纺　滕云飞

交织　上海南外滩地块城市设计

陈宏　魏崴　[美]古斯塔夫·克伦贝尔(Gustavo Crembil)　编著
*
中国建筑工业出版社出版、发行（北京西郊百万庄）
各地新华书店、建筑书店经销
上海盛通时代印刷有限公司制版印刷
*
开本：889×1194毫米　1/20　印张：15　字数：735千字
2013年11月第一版　2013年11月第一次印刷
定价：128.00元
ISBN 978-7-112-16071-6
　　　（24830）

版权所有　翻印必究
如有印装质量问题，可寄本社退换
（邮政编码　100037）